Strategizing Continuous Delivery in the Cloud

Implement continuous delivery using modern cloud-native technology

Garima Bajpai

Thomas Schuetz

BIRMINGHAM—MUMBAI

Strategizing Continuous Delivery in the Cloud

Group Product Manager: Preet Ahuja

Publishing Product Manager: Vidhi Vashisth

Senior Editor: Runcil Rebello

Technical Editor: Arjun Varma

Copy Editor: Safis Editing

Project Coordinator: Ashwin Kharwa

Proofreader: Safis Editing

Indexer: Rekha Nair

Production Designer: Joshua Misquitta

Marketing Coordinator: Rohan Dobhal

First published: August 2023
Production reference: 1200723

Published by Packt Publishing Ltd.
Grosvenor House
11 St Paul's Square
Birmingham
B3 1RB

ISBN 978-1-83763-753-9

www.packtpub.com

I would like to thank Canada DevOps Community of Practice, where I have been able to brainstorm a lot of ideas with chapter leads and senior community members. I would also like to thank the open source community, which provides us with a platform to steer thought leadership in the DevOps and continuous delivery space, especially the Continuous Delivery Foundation ambassadors. My gratitude toward my co-author, Thomas, and the entire team of Packt; without them, I would not have been able to bring this book to life. Lastly, I would like to express my gratitude toward Amit Kumar Mishra and my daughter Akira for their relentless support and encouragement.

– Garima Bajpai

First and foremost, I would like to thank my loving partner, Jacqueline, for her patience while writing this book and her support. Second, I would like to thank all of the people who supported my journey to the cloud in the last few years. Last but not least, I also want to thank my students for asking challenging questions that found their way into this book and those people who reviewed and gave guidance for this book.

– Thomas Schuetz

Contributors

About the authors

Garima Bajpai is a thought leader and industry leader in DevOps and cloud technologies. She is the founder of DevOps Community of Practice, Canada. She leads the ambassador program for the Continuous Delivery Foundation. Some might know her as a course contributor and instructor for various DevOps courses from leading certification bodies. She has over two decades of experience leading large-scale R&D with a variety of different teams and has helped them adapt DevOps to be able to increase team productivity when it comes to cloud resource deployment. Furthermore, she has collaborated on and contributed to giving many international conference talks, written several technical blog posts, and published white papers.

Thomas Schuetz is a cloud-native engineer, advocate, and educator living in Austria with a strong technical background in systems engineering. He is specialized in cloud-native continuous delivery as well as infrastructure-as-code, is a Cloud Native Computing Foundation and Continuous Delivery Foundation Ambassador, as well as being the maintainer of some open source projects, such as Keptn and K8sGPT. Besides that, he shares his experience with students at the University of Applied Sciences Burgenland, Austria.

About the reviewers

Nidio Dolfini has majored in SI and has a post-graduate degree in the cloud, with more than 15 years of experience in the IT area, in which he has worked at educational and technological companies. For about seven years, he was an IT coordinator, obtaining leadership, management, and communication skills, which allowed him to develop a strategic vision and a results-oriented approach. It was then that he decided to swap the management part for the technical, and today, he currently works as a DevOps engineer and a cloud professor. He always keeps up to date with the current trends and best practices. His focus is to aid his team in building and maintaining a reliable and scalable infrastructure and always bringing innovation.

I would like to thank Naval, Eduardo Pato, Felgas, and Ferrazani for believing in me and for the opportunity given. Dudu, Dirceu, and Herbert, for always supporting me throughout my ventures. Ana, for her patience throughout the times I had to study and prepare this book review. Bia and Sophia for loving me regardless of how far they are, and Teacher Bruna for helping me with my English course.

Antonio Masucci (Anto) is an experienced software engineer known for his expertise in DevOps and platform engineering. With over a decade of experience, he began his career as a backend engineer before transitioning into roles focused on DevOps and continuous delivery.

Passionate about sharing knowledge, Anto actively engages with the continuous delivery and platform engineering communities through his popular YouTube channel, *OutOfDevOps*. Through this platform, he fosters discussions on the latest industry trends and practices.

Throughout his professional journey, Anto has led platform teams and spearheaded digital transformations in various industries, including gaming, publishing, automotive, and finance.

I would like to express my deepest gratitude to my loving wife and my wonderful children for their unwavering support throughout my journey. Their constant encouragement, understanding, and love have been my pillars of strength, fueling my passion and drive to succeed. I am forever grateful for their presence in my life and the countless sacrifices they have made to uplift and inspire me. I am truly blessed to have them by my side.

Table of Contents

Part 1: Foundation and Preparation for Continuous Delivery in the Cloud

1

2

3

Creating a Successful Strategy and Preparing for Continuous Delivery   29

4

Setting Up and Scaling Continuous Delivery in the Cloud   45

Part 2: Implementing Continuous Delivery

5

6

7

8

Security in Continuous Delivery and Testing Your Deployment   111

Part 3: Best Practices and the Way Ahead

9

Best Practices and References   127

Preface

Many organizations are embracing cloud technology to remain competitive. Implementing and adopting your software development processes, tools, and techniques while modernizing a cloud-based ecosystem can be challenging. *Strategizing Continuous Delivery in the Cloud* will help you to understand why you need to modernize continuous delivery, a core component of software development and delivery, and how to drive the convergence of infrastructure and software application deployment in the cloud.

You will begin by discovering how continuous delivery in the cloud is different from traditional approaches to continuous delivery and will look at the steps to build the right strategy that works best for your organization. Further, you will get guidance on developing, testing, integrating, deploying, and operating cloud-native software in different types of cloud-based environments. The book covers in detail the prerequisites for onboarding continuous delivery in the cloud from an organizational and technical perspective. You will delve into key aspects of organizational and technological readiness to lay out successful strategies for overcoming core challenges on your cloud journey. Finally, you will learn about the dos and don'ts to stay on top of your DevOps game long after the starting point.

By the end of this book, you will have a solid understanding of how to choose the right type of cloud environment and the right technologies for modernizing continuous delivery.

Who this book is for

This book is targeted toward operations engineers, site reliability engineers, DevOps architects, and engineers who are strategizing, planning, and implementing continuous delivery in the cloud. You are expected to have a basic understanding of CI/CD concepts and be familiar with the cloud ecosystem, DevOps, and CI/CD pipelines.

What this book covers

Chapter 1, Planning for Continuous Delivery in the Cloud, provides a brief introduction to continuous delivery concepts, its relevance, and its benefits. In addition, we will also describe the step-by-step implementation of continuous delivery with the help of some tools and techniques used industry-wide.

Chapter 2, Understanding Cloud Delivery Models, describes the relevant basics and characteristics of cloud computing (e.g., delivery models). This chapter should help you to understand which things are useful when dealing with continuous delivery in the cloud.

Chapter 3, Creating a Successful Strategy and Preparing for Continuous Delivery, covers in detail the prerequisites for onboarding continuous delivery in the cloud from an organizational perspective. It presents an introduction to various aspects of organizational readiness and explains how to lay out successful strategies for continuous delivery in the cloud. The chapter provides guidance on implementing, managing, and improving continuous delivery in the cloud. It also digs deeper into preparing for the modernization needs of CD with respect to the cloud, core challenges, aligning and connecting a roadmap with the cloud to maximize the outcome, and lastly, competence and upskilling needs.

Chapter 4, Setting Up and Scaling Continuous Delivery in the Cloud, provides a detailed view of setting up continuous delivery in the cloud to maximize the overall return on investment of software development, infrastructure, and management. The chapter provides an overview of scaling up continuous delivery in the cloud for enterprises. It includes organizational characteristics, requirements for scaling CI/CD, and ownership of CI/CD with roles and responsibilities.

Chapter 5, Finding Your Technical Strategy Toward Continuous Delivery in the Cloud, describes the architectural decisions and considerations that must be taken when aiming for continuous delivery. Furthermore, it sheds some light on topics that could put a CD environment and strategy at risk.

Chapter 6, Achieving Successful Implementation with Supporting Technology, describes some additional technical measures for your CD strategy and how they might help to achieve a successful implementation.

Chapter 7, Aiming for Velocity and Reducing Delivery Risks, describes why lead times in application delivery have an impact, why they could be hard for developers, and how we can measure performance.

Chapter 8, Security in Continuous Delivery and Testing your Deployment, describes why security in continuous delivery is important. Furthermore, it deals with security in the software supply chain and which checks can be done to ensure that the delivered software is secure.

Chapter 9, Best Practices and References, focuses on industry best practices and a few case studies as a reference.

Chapter 10, Future Trends of Continuous Delivery, focuses on the key trends of continuous delivery and connected advancements. We focus on the mid to long term to introduce readers to evolutionary changes and key considerations for their adoption.

Chapter 11, Contributing to the Open Source Ecosystem, focuses on open source projects for potential contributors. We introduce projects and ways to contribute to the open source ecosystem for further reading.

Chapter 12, Practical Assignments, focuses on practical assignments to test your knowledge.

To get the most out of this book

You'll need to have introductory knowledge about software development, continuous delivery practices, and cloud technologies; a basic understanding of CI/CD concepts; familiarity with the cloud ecosystem, DevOps, and CI/CD pipelines; basic knowledge of continuous integration and delivery tools and the cloud ecosystem; and use of at least one cloud provider, for example, AWS or Azure.

Software/hardware covered in the book	Operating system requirements
Kubernetes, Helm, Kustomize	Windows,MacOs and Linux

Download the example code files

We have code bundles from our rich catalog of books and videos available at `https://github.com/PacktPublishing/`. Check them out!

Conventions used

There are a number of text conventions used throughout this book.

`Code in text`: Indicates code words in text, database table names, folder names, filenames, file extensions, pathnames, dummy URLs, user input, and Twitter handles. Here is an example: "The controller will also ensure that the state defined in the object is met, so if you change it to `expose=false`, the controller will delete the service."

A block of code is set as follows:

```
apiVersion: apps/v1
kind: Deployment
metadata:
  name: my-cloudy-app
spec:
  replicas: 3
```

When we wish to draw your attention to a particular part of a code block, the relevant lines or items are set in bold:

```
apiVersion: apps/v1
kind: Deployment
metadata:
  name: my-cloudy-app
spec:
  replicas: 3
vm.dirty_expire_centisecs = 3000
```

Any command-line input or output is written as follows:

```
mkfs.ext4 -E stride=16,stripe-width=160 /dev/sdX
```

Bold: Indicates a new term, an important word, or words that you see onscreen. For instance, words in menus or dialog boxes appear in **bold**. Here is an example: "Select **System info** from the **Administration** panel."

> **Tips or important notes**
> Appear like this.

Get in touch

Feedback from our readers is always welcome.

General feedback: If you have questions about any aspect of this book, email us at customercare@packtpub.com and mention the book title in the subject of your message.

Errata: Although we have taken every care to ensure the accuracy of our content, mistakes do happen. If you have found a mistake in this book, we would be grateful if you would report this to us. Please visit www.packtpub.com/support/errata and fill in the form.

Piracy: If you come across any illegal copies of our works in any form on the internet, we would be grateful if you would provide us with the location address or website name. Please contact us at copyright@packt.com with a link to the material.

If you are interested in becoming an author: If there is a topic that you have expertise in and you are interested in either writing or contributing to a book, please visit authors.packtpub.com.

Share your thoughts

Once you've read *Strategizing Continuous Delivery in the Cloud*, we'd love to hear your thoughts! Scan the QR code below to go straight to the Amazon review page for this book and share your feedback.

https://packt.link/r/1837637539

Your review is important to us and the tech community and will help us make sure we're delivering excellent quality content.

Download a free PDF copy of this book

Thanks for purchasing this book!

Do you like to read on the go but are unable to carry your print books everywhere?

Is your eBook purchase not compatible with the device of your choice?

Don't worry, now with every Packt book you get a DRM-free PDF version of that book at no cost.

Read anywhere, any place, on any device. Search, copy, and paste code from your favorite technical books directly into your application.

The perks don't stop there, you can get exclusive access to discounts, newsletters, and great free content in your inbox daily

Follow these simple steps to get the benefits:

1. Scan the QR code or visit the link below

https://packt.link/free-ebook/9781837637539

2. Submit your proof of purchase
3. That's it! We'll send your free PDF and other benefits to your email directly

Part 1: Foundation and Preparation for Continuous Delivery in the Cloud

This part focuses on continuous delivery concepts and their relevance and benefits. Furthermore, the characteristics and usages of cloud systems in the context of continuous delivery are elaborated upon.

This part contains the following chapters:

- *Chapter 1, Planning for Continuous Delivery in the Cloud*
- *Chapter 2, Understanding Cloud Delivery Models*
- *Chapter 3, Creating a Successful Strategy and Preparing for Continuous Delivery*
- *Chapter 4, Setting Up and Scaling Continuous Delivery in the Cloud*

1

Planning for Continuous Delivery in the Cloud

This chapter provides a brief introduction to **continuous delivery** (**CD**) concepts, their relevance, and their benefits. In addition, we will also describe the step-by-step implementation of CD with the help of some industry-wide used tools and techniques, as follows:

- Understanding the CD ecosystem
- Key benefits of CD
- From **continuous integration** (**CI**) to CD
- Progressive delivery
- Cloud-based implementations of CD

Understanding the CD ecosystem

This chapter is intended to reflect the recent development and evolution of the software delivery approach, mostly focusing on changes to CD and connected advancements underpinned by the emergence of CI technologies, new tools, processes, practices, and management structures. To understand the advancements of CD in the cloud better, it is important to connect the dots and learn about internal and external factors influencing the advancement of CD. The term CD has been around for a long time; yet, let us start from the basics. The most widely referred definition of CD is provided by Jez Humble (and can be found at `https://continuousdelivery.com/`):

> *"Continuous Delivery is the ability to get the changes of all types of new features, config changes, bug fixes, and experiments into production or into the hands of the user safely and quickly in a sustainable way."*

With advancements in CD, there are several attempts to redefine or refine the definition of CD. For example, AWS (at `https://aws.amazon.com/devops/continuous-delivery/`) defines CD as follows:

"Continuous delivery is a software development practice where code changes are automatically prepared for a release to production."

According to another definition by Azure (at `https://docs.microsoft.com/en-us/devops/deliver/what-is-continuous-delivery`), CD is defined as follows:

"Continuous delivery (CD) is the process of automating build, test, configuration, and deployment from a build to a production environment."

To understand it from Google's perspective (at `https://cloud.google.com/solutions/continuous-delivery`), it is simply the following:

"End-to-end automation from source to production."

Although there are variances and differences in definitions of CD, it makes it easier for us to reflect that CD is evolving and is defined by a set of characteristics that work together in the ecosystem.

As we move ahead in this book, we will describe the building blocks of CD and their relationship.

Defining the CD ecosystem

For now, let's try to understand the factors influencing the evolution of CD to define the ecosystem:

- **Speed**: It is evident that most of the definitions of CD lead to automation, making it visible that speed is one of the core drivers for the evolution of the CD and its connected ecosystem. Many key advancements are linked to removing the bottleneck to speed and lead time: starting from the evolution of practices such as CI/CD to the explosion of cloud-based pre-built pipelines, as well as including automation in build-test-deploy practices.

- **Security**: The fundamental shift toward securing software and addressing key risks with the utmost priority has led to the systematic evolution of the CD ecosystem with new tools and techniques, including security from the inception stages of CD.

- **Software development itself**: Developers are inclined to move away from traditional approaches to software development to more modern ways with the integration of Agile methodology, making use of more progressive, efficient tools and processes to eliminate lead times and quality issues. One example of such an application is Azure Boards, which provides a choice of Agile planning tools, many of which work in combination with CD practices.

- **Software life cycle management**: With velocity, scale, interdependence, and the growing complexity of the software landscape comes the need to simplify software life cycle management. Quick discovery and fixing of software bugs, accelerating software updates, comprehensive assessment of dependencies, and razor-sharp focus on vulnerability management are essential elements.

- **Software operations**: Finally, the **return on investment** (**ROI**) on software is realized in production. DevOps, SRE, and the emergence of cloud-based services are revolutionizing our CD ecosystem with a focus on building cross-functional teams, platforms, and tools.

The preceding factors indicate that the CD ecosystem comprises dynamic internal and external components. Taking inspiration from a natural ecosystem, if we attempt to define the CD ecosystem, we come up with this:

The CD ecosystem is a distributed, adaptive, and techno-functional system of CD practices, processes, tools, and techniques that support the evolution of CD.

Characteristics of the CD ecosystem

In order to be able to deliver continuously, we need a system of practices, processes, tools, and techniques that are reliable and can produce repeatable results consistently. Some of the basic ingredients for providing such outcomes are the following:

- **Self-organizing and adaptive**: The CD ecosystem, as with any other such system, must be in a state of equilibrium. No single specific outcome can drive the construct and choice of practices, tools, or techniques entirely. For example, speed is one driver for CD, but the ecosystem must have components that keep the ecosystem close to equilibrium by including components/tools through which dynamics of security, speed, quality, and value are presented in a balanced way.

- **Dynamic**: The CD ecosystem consists of evolving tools and practices and integrates with emerging technology, so it makes sense that the CD ecosystem is dynamic. For example, the evolution of Kubernetes from an internal container ecosystem at Google managed by a community of contributors into an enterprise-grade de facto orchestration engine, with its adoption by cloud providers, adds an array of new features to CD. Interoperability and event-driven architecture evolution are key areas of progress to emphasize the dynamic characteristic of the CD ecosystem.

- **Distributed**: The CD ecosystem should be able to provide our developers and practitioners with an array of tools and services. It connects directly to the evolution of the marketplace approach. For example, cloud service providers such as Azure enable you to build and deploy applications on an array of platforms and languages. Azure also offers two options for version control—Git and Azure Repos, and many more such choices.

- **Supports an "as-a-service" model**: With time, the key components of the CD ecosystem will be degraded or replaced by other new and advanced features. We must focus on a consumption-based model for features and measure their usage over time. One of the key drivers of an **as-a-service** model is that it helps to build a self-regulating and self-sustaining ecosystem. Many services from cloud-based CI/CD tools and application vendors have community versions and tiered subscriptions for on-premises, hybrid, or enterprise versions, supporting a pay-per-use pricing model.

Key components of the CD ecosystem

For any ecosystem to thrive, the key components must act together to advance it by catering to evolving demands. The CD ecosystem will continue to evolve if progressive tools and applications continue to challenge the status quo – right from basic CI/CD tools and applications for version control and workflow automation through to specialized cloud-native CI/CD tools that can be run on any cloud and build a strong progressive foundation anywhere.

The CD ecosystem comprises three key components that continuously steer the evolution of the ecosystem by delivering value within and outside the boundaries of the ecosystem CI/CD tools and applications: consumers, producers, and practitioners. Let us look at them in detail:

- **The role of CI/CD tools and applications as a consumer**: The increasing adoption of digital technologies creates huge market potential for CD. If CD provides a competitive advantage to organizations by delivering software features frequently reliably and cost-effectively, the ecosystem is bound to flourish. The role of consumers in the ecosystem is to provide feedback to CI/CD tools and application vendors and continue to experiment and modernize the CI/CD pipeline with progressive features, tools, and applications to help unlock new revenue streams.

- **The role of CI/CD tools and applications as producers**: The CI/CD ecosystem needs investment and integration with emerging technologies to facilitate a competitive advantage for consumers. The tools and applications that are the foundation of the CD ecosystem should also be the consumers of the ecosystem. For example, CI/CD tools and applications are often architected with microservices and rely on container-based deployment. They are designed to support interoperability and integrate well with other cloud-native tools and cloud-based services.

- **The role of CI/CD practitioners as an actor in an ecosystem**: The practitioners are the final actors in the ecosystem through which value is realized. The practitioners will churn out the waste in the CI/CD pipeline. Through the adoption of various tools and applications, practitioners will continuously optimize the ecosystem, and create perspectives on new applications, tools, and features. For example, pushing toward more interoperability standards with the rise of hybrid pipelines, the adoption of co-pilot features, and a more platform-centric approach toward CD to deliver software features seamlessly is paving the way for new developments.

Managing the CD ecosystem

Several organizations have already embarked on their CD journeys. With that, the need to effectively manage investments, develop and improve the current technical adoption, and steer strategic outcomes through CD becomes an urgent priority.

Many industry reports highlight that the full potential of CI/CD and the connected ecosystem is yet to be realized. There are multiple reasons associated with this, such as siloed initiatives, ad hoc investment, and lack of upskilling. These factors indicate the need to manage the CI/CD landscape with a multidimensional outlook.

The management of CD also must be structured yet agile in a way. Planning, organizing, implementing, and monitoring the outcome of CD must be modern and fit for purpose. To avoid delays, bottlenecks, and constraints, the management process should be tailored to CD.

In this subsection, we will present some guidance on effectively managing CD. We will start with a CD reference architecture, further elaborate on the role of CD frameworks, and eventually highlight the need to have a CD **Community of Practices (CoPs)**. In subsequent chapters, we will detail these aspects along with other important dimensions of creating and implementing CD in the cloud strategically:

- **CD reference architecture**: This refers to the abstract concepts that can outline a structure or a construct for CD:

  - Provides a common vocabulary

  - Provides guidance about the functions and their interaction (e.g., APIs)

  - Can be defined at a different level of abstraction

  - Can be instantiated for a particular domain or specific projects

  - Can be mapped to specific sets of goals, as no single architecture can fulfill all needs

  An example of a reference architecture and associated service offering in the cloud is Open Source on AWS: `https://aws.amazon.com/blogs/devops/choosing-a-well-architected-ci-cd-approach-open-source-on-aws/`.

  Another example of a CD reference architecture is one being worked on by the **Continuous Delivery Foundation (CDF)**:

- **A CD framework**: A **framework** is a pre-built general or special-purpose architecture that's designed to be extended for a set of consumers, such as the healthcare industry, the defense sector, or the financial community. CD tools, along with cloud technologies, are constantly undergoing changes to improve the software development environment and practices. It is hard to keep pace with upskilling without structured support from the community of practitioners.

  As an example of a community of practice, here are a collection of technical articles and blogs published or curated by Google Cloud Developer Advocates: `https://medium.com/google-cloud`.

The AWS Community Builders program is one of the unique programs launched by AWS for the community of practitioners to share resources and educational content and build a networking ecosystem to support the community.

Other communities provide similar guidance: the DevOps Institute's *DevOps in the Wild* community and the CDF community for open source CD tools and best practices, to name two.

So far, we have introduced some terminology we will use throughout the book, and this section has ensured that you have a good understanding of the main concepts of CD and its key components. We also touched upon the aspects of CD in the cloud, which will be illustrated in the following chapters in detail.

Key benefits of CD

In this section, we will look at the benefits of CD in more detail. Building upon the idea of the key characteristics of CD being speed, enabling security and progressive practices for software development, and operation and life cycle management, let's look at the benefits.

There are some industry references. **DevOps Research And Assessment (DORA)** is an initiative by Google to support organizations to achieve high performance by embracing DevOps according to actionable guidance, DORA metrics (`https://cloud.google.com/blog/products/devops-sre/using-the-four-keys-to-measure-your-devops-performance`), which can give you a measure of the positive business outcomes from adopting CD. For us, it will be easier to map the key characteristics to the business outcomes and key benefits of CD:

Characteristics of CD	Description	Business outcome
Speed	How frequently can we release?	% gains on market shares from launching new products and services
Security	How trustworthy are we?	% reduction of the cost of risk mitigation and cost avoidance of security breaches
Agility in software development	How fast can we respond to the changing needs of the customer?	% revenue gains on the accelerated time to market for changes and new products
Simplifying software life cycle management	How resilient are we?	% gains from the cost reduction of application failures
Collaboration with software operations	How effectively do we communicate and share information between teams?	% gains from enhanced team productivity

Table 1.1 — Mapping CD characteristics to key benefits

Let us now look at the benefits of CD in the cloud.

Benefits of CD in the cloud

Taking one step forward, cloud service providers combined the benefits with the service offerings to facilitate better business outcomes:

Characteristics of CD	Business outcome	Service offering of cloud service providers
Speed	% gains on market share from launching new products and services	Automate the software delivery process through an already tested CI/CD pipeline. Improve developer productivity by removing mundane tasks.
Agility in software development	% revenue gains on accelerated time to market for changes and new products	Feature development in increments, which are discrete and small so they can be delivered with agility.
Security	% reduction of the cost of a long list of risk mitigation and cost avoidance strategies aimed at security breaches	Cloud providers offer a marketplace of security tools, which can be easily integrated with the pre-built pipelines through APIs.
Simplifying software life cycle management	% gains from the cost reduction of application failures	With managed services from cloud providers, it is easier to roll out updates. What used to take weeks and months can now be done in days or even hours.
Collaboration with software operations	% gains from enhanced team productivity	Implementing practices adopted by several organizations. Example: AWS has a number of certified DevOps Partners who can provide resources and tooling.

Table 1.2 — Mapping CD's benefits to cloud service provider offering

In this section, you read a short overview of the CD ecosystem and how it can help you in achieving your goals. In the next sections, you will be introduced to the differences between CI and CD and where CD in the cloud can support you.

From CI to CD

When delivering software to our customers, the customer assumes that it runs reliably and without significant problems. Any misfunction or outage of the software—depending on its purpose—might lead to a loss of reputation on the customer side and the people and company that developed the software. Think of an issue where code was written weeks or months ago, the software is already deployed in the customer's environment, and it crashes randomly after some time. While troubleshooting, we might tackle the following issues:

- **Time pressure**: As the software has already been delivered to production and customers have been affected by its misbehavior, it is obvious to provide a resolution during this process. As people providing this might work under pressure, it's not unlikely that new issues might be introduced.

- **Determine which change introduced the misbehavior**: In our case, the change might have been introduced a long time ago, so one of our major challenges might be finding out when or in which context the change has been introduced.

- **Finding the needle in the haystack**: Our software might consist of lots of components. If we can't find out which change introduced the issue, we might want to find where in the code it happens.

- **Dev/prod parity**: As our working copy of the code might already have progressed further than the version we are fixing the issue for, we will also have to update newer versions of the software.

These are only examples of problems we might face while fixing such issues. To tackle such cases, we want to find problems in our software very early and build automation around building, testing, and publishing artifacts, called **CI**. The process of CI should ensure that the code checked in is continuously built and tested. Therefore, problems are found at a very early stage.

Using CI and CD methods, we can find problems early and avoid them hitting the customer's system. Before we dive deeper into CD processes and implementation, we will take a brief look at the building blocks of a typical CI/CD infrastructure:

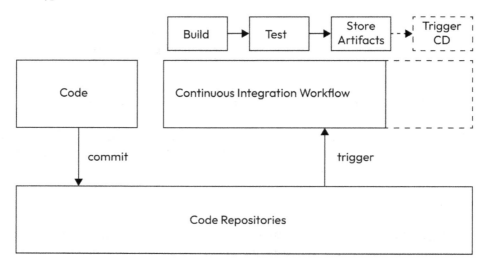

Figure 1.1 – Building blocks of a CD infrastructure

In current development setups, developers store their code in **source code management** (**SCM**) systems, such as Git, Mercurial, or Subversion. CI systems, such as Jenkins or Tekton, can either watch these repositories for changes on specific branches or tags or get triggered by the SCM systems to initiate workflows. These workflows may include static code analysis steps, such as linting and static security checks, before building, but also more dynamic tests, which can only be done against the running system, such as API tests or Dynamic Application Security tests.

Usually, these steps run on every commit in so-called feature branches. At some point in time, a developer might decide that a developed code part (**feature**) might be finished and therefore include it in the shared code base and raise a pull or merge request. As a best practice, there might be code reviews for this integration step, which can be automatic or manual. Humans will do the manual part. Furthermore, automatic checks can be done by a CI workflow, which should ensure that the code checked into the code repository is not only compilable but also stable and, in the best case, secure.

Depending on the release strategy, a successful merge to a central shared repository could trigger the CD process. This could also be done when the code is merged into specific branches or with defined tags. As discussed in later chapters of this book, the code will be deployed in various environments (**stages**) during the further CD process, and tests should ensure that the software is in a deployable state at the end of the process, and that deliverable artifacts are available. Although the terms CI and CD are often used interchangeably, the process of deploying automatically to the production systems is called **CD**.

In this section, we learned that it is important to find problems in software early by building and testing software continuously. Furthermore, CD helps us keep the software in a deployable state at any time. Last but not least, CD can help us in accelerating our time to market by deploying to production automatically.

Progressive delivery

Many issues could come up with continuously deploying to production. Even if the software is tested well during the previous phases and our deployment mechanisms are working perfectly, we can encounter situations where everything is running from a technical perspective and still the experience for the end users is very dissatisfying. For instance, a customer might click on a button on a website, and it takes about 30 seconds to get a response, although there is not much load on the system. Such things could come up because the software is not tested under the conditions present in the production environment, and the things users are executing have not been tested in this constellation before. Long story short, we should always aim to deploy our software often and automatically, but we need additional mechanisms that safeguard our deployment process. One of them is progressive delivery.

"Progressive delivery is the process of pushing changes to a product iteratively, first to a small audience and then to increasingly larger audiences to maintain quality control."

(Orit Golowinski, `https://www.devopsinstitute.com/progressive-delivery-7-methods/`)

Using progressive delivery, the software is deployed in a phased way, ensuring that not all users are affected by a problem. In *Chapter 6*, we will take a closer look at Blue-Green Deployments and Canary Releases, which can be used to not only create preview environments but also to shift production traffic between different versions of a product. At this point, we are able to deploy automatically after a new feature has been introduced and might be able to deliver it to a small user base to see how it performs. As we aim for automation and might not want to make such decisions manually, modern progressive delivery controllers are able to make such decisions based on data. For instance, a new software version gets automatically deployed to 1 of 5 production environments. We want to ensure that requests do not take longer than 500 milliseconds, and in the newer release, the response time must not be more than 10% higher than before. Therefore, we monitor such parameters and configure our progressive delivery controller to wait for some time and shift the next 20% of traffic if our conditions are met.

Progressive delivery is not only used to limit the blast radius of errors but also to validate user acceptance of newly implemented features. Using observability data, we could also find out whether there is a decrease in orders in various timeframes when a new feature is rolled out and can automatically roll back the feature in this case.

We will dive deeper into the various deployment strategies, as well as observability and feature flagging, in *Chapter 6* and *Chapter 7*.

Cloud-based implementations of CD

All of the things we discussed in the previous sections could be implemented in your own data center. The code could be hosted in a repository, a local CI server might watch for changes, and CD tooling could deploy the applications to the production environments. In the rest of the book, however, we will mainly cover *CD in the cloud*. Therefore, we assume that the software is built, deployed, and operated in cloud environments. We will go more into detail on cloud characteristics and delivery models in *Chapter 2*. In this section, we will deal with a few technologies, as well as the benefits and drawbacks when doing CD in the cloud.

When dealing with CD and deployment, we will often refer to cloud-native applications and technologies, which are defined as follows.

Cloud-native technologies empower organizations to build and run scalable applications in modern, dynamic environments such as public, private, and hybrid clouds. Containers, service meshes, microservices, immutable infrastructure, and declarative APIs exemplify this approach.

These techniques enable loosely coupled systems that are resilient, manageable, and observable. Combined with robust automation, they allow engineers to make high-impact changes frequently and predictably with minimal toil (the *Cloud Native Computing Foundation* – `https://www.cncf.io/about/who-we-are/`).

As defined previously, many technologies and products may be involved when delivering cloud-native systems. We will often hear three terms: virtualization, containers, and container orchestration/ Kubernetes. Let us look at them in some detail:

- **Virtualization** is one of the core techniques in cloud computing and allows more efficient usage of resources and supports the implementation of typical cloud characteristics; for example, on-demand self-service, resource pooling, and rapid elasticity. Using virtualization allows the abstraction of software from hardware; typically, virtualization is referred to the usage of virtual machines, where compute resources of one physical host are *partitioned* into multiple virtual machines.

- **Containers** provide a way to package the runtime environment, as well as the application code, into a single artifact that can run on a container runtime. While containers have existed for a long time in the industry (OpenVZ and BSD jails), Docker extended the technology to build containers in a systematic way and to easily add containerization to the development process. While virtual machines introduce some overhead into the system, as they mostly run their own kernel, containers run with almost no overhead, which leads to much better resource usage and a higher possible load of workloads on one machine.

- **Container orchestration** deals with many containers in many physical environments and ensures that container workloads are not only scheduled on the right node but also get the resources they need. At the time of writing, Kubernetes is the de facto standard for container orchestration. One of the main benefits of using a container orchestration platform is the simple fault tolerance and scaling of applications. The scheduler of the platform takes care of the desired number of containers and the user's needs, and—if the resources are available— manages them according to the configured specifications. Furthermore, the usage of such an infrastructure enforces the operators to use configurations, which may either be in the format provided by the platform or in a framework for managing deployments for this platform. As a result, the used infrastructure is documented in code and can be recreated and duplicated with little effort. Orchestration platforms generally provide load-balancing mechanisms and the possibility to define health and readiness checks, which makes it easy for users to build auto-scaling applications.

When deploying to cloud-based systems, we also must think about the infrastructure. To define the infrastructure, the term **Infrastructure as Code** has emerged, which is the practice of describing the target state of the infrastructure in *declarative language*. One of the major challenges we will also cover in this book is the convergence of the infrastructure and the application deployment.

Every major cloud service provider provides a framework for deploying applications automatically to their environment. There are Infrastructure-as-Code modules available for various toolsets and proprietary solutions. As a further consequence, each cloud provider provides a managed code repository and CI tooling to build software from code and container registries to store the container images. Last but not least, there are many possibilities for running containers/applications in the cloud providers, such as a managed Kubernetes service:

Service	AWS	GCP	Azure
Code Repository (Git)	Code Commit	Cloud Source	Azure Repos
CI Pipeline	Code Pipeline	Cloud Build	Azure Pipelines
Container Registry	Elastic Container Registry	Container Registry	Azure Container Registry
Managed Kubernetes	**Elastic Kubernetes Service (EKS)**	**Google Kubernetes Engine (GKE)**	**Azure Kubernetes Service (AKS)**

Table 1.3 — Examples of cloud services for CD

In this section, we discussed a few basics of cloud-based CD implementation and services, which can be consumed through cloud providers. As we progress through this book, we will use the resources provided by these cloud providers, as well as tools that can be installed on cloud provider resources.

Summary

In this chapter, we introduced the foundational aspects of CD, starting with its definition and characteristics, and explaining the CD ecosystem at a high level.

Once we learned about all the benefits of CD, we took a closer look at the technical side and its prerequisites and learned that CI helps us build and test software continuously. Furthermore, CD ensures that our software is in a deployable state at any time, and Continuous Deployment, which is often used interchangeably with CD automates the deployment of our application to production systems.

As failures in automated deployments could have a very large blast radius, progressive delivery can reduce this by gradually deploying the application to a limited number of users. To also automate the evaluation of the behavior of the systems, we might rely on data from our observability solution to make decisions if the criteria to shift more traffic to the new version are met. Last but not least, we can also use these mechanisms to assess the business impact of a new feature and take action if this leads to, for example, a loss of revenue.

In the final part of this chapter, we started to deal with cloud-specifics related to CD that we will need in future chapters of the book. Therefore, we introduced and explained the terms **Virtualization**, **Containers**, and **Container Orchestration** and learned that we need to take care of our infrastructure (using Infrastructure as Code) before deploying applications on it. Finally, we discovered that there are many cloud services that will help us on our journey to CD in the cloud.

Throughout the next chapter, we will go deeper into the characteristics of the cloud and its delivery models. Furthermore, you will learn why delivering to the cloud is different from traditional environments.

Further reading

- The book *Continuous Delivery: Reliable Software Releases Through Build, Test, and Deployment Automation* by Jez Humble and David Farley

2

Understanding Cloud Delivery Models

In the previous chapter, we introduced some core concepts of continuous delivery and gained insights into cloud-based implementations of it. Although we learned that there are tools in the cloud that might help us implement continuous delivery, we will briefly touch upon the characteristics of the cloud itself and how it can help us deliver our software more efficiently and conveniently. But before we do that, we'll take a brief detour into how cloud computing changed the build/**continuous integration (CI)** processes in the last few years.

When companies started using CI they often started with one instance of their CI/automation server (such as Jenkins) and ran their builds directly on these servers. After some time, they might have discovered that these builds ran better in parallel and added some (more or less static) runners on virtual machines or bare metal to handle this load. As you might imagine, these pre-provisioned runners have their weaknesses. For instance, they might be overloaded in peak times but might be idle on weekends or during the night. Furthermore, these build servers and runners were often maintained *by hand*, and reconfiguring or reinstalling these environments sometimes became very hard. In addition, changes to this infrastructure were maintained on wiki pages and therefore had to be updated manually each time the infrastructure changed to keep it up to date.

With the rise of cloud services and containerization, many of these problems got addressed more or less by themselves. A continuous integration server needn't necessarily be created by hand, and runners can be started and stopped on demand. In this chapter, we will discuss how some characteristics of the cloud helped us get there. But before that, let's have a look at the list of topics we will cover:

- Cloud deployment models
- Cloud characteristics
- Cloud service models
- Why is continuous delivery different in the cloud?
- Opportunities and risks of practicing cloud-based continuous delivery

Cloud deployment models

When we started to deal with cloud computing, the term *cloud* was often referred to as *must be running on AWS, Azure, Google Cloud Platform, or a similar cloud provider*. There are multiple possibilities to run cloud services in public clouds (such as AWS, GCP, or Azure), private clouds (on- or off-premises), or hybrid clouds.

Before going further, let's briefly discuss the different deployment models and their characteristics. **Cloud deployment** refers to the ownership, management, and evolution of cloud services. One common way to categorize cloud deployments is by the deployment models and their characteristics. The following table describes the different types of deployment models and the mapping of their characteristics in general:

Deployment Model	Description	Characteristics
Public cloud	The public cloud makes the cloud services offered to the public available via the public internet, and the services can be pooled with different organizations.	The key characteristics of the public cloud contain the cloud characteristics described in *Table 2.2* of this chapter, with one differentiating characteristic of pay-as-you-go or subscription-based services
Private cloud	The private cloud makes cloud-based services available to a single organization. The data center is either owned and managed by the organization or a third party.	The key characteristics of the private cloud contain the cloud characteristics described in the *Cloud deployment models* section of this chapter, with the differentiation that the hardware and operations are owned by the organization, with resource pooling within the organization's business units
Hybrid cloud	The hybrid cloud makes cloud-based services available to a single organization with some services hosted in the public cloud. The organization can move workloads between public and private clouds in a controlled way.	The key characteristics of a hybrid cloud contain the cloud characteristics described in the *Cloud deployment models* section of this chapter, with the differentiation that it is possible to use some public cloud services, which are subscription-based, and move the workloads between public and private clouds for specific workloads

Table 2.1 – Cloud deployment models and their characteristics

In a typical cloud setup, we could have a hybrid, public, or private deployment model. Going back to our CI server example, we might run our CI server in our data center while it's connected to a container orchestrator, which is used to schedule the CI jobs in a private setup. This would help our company scale the process and keep information private. In our private data center, we should not expect indefinite resources or bandwidth.

It might also be an option to run the CI server and git repositories in our data center, connect it to a public cloud provider, and only run the executors in the public part of the cloud on demand. Since we have resources mixed into our private cloud but also on the provider infrastructure, this can be referred to as a hybrid cloud infrastructure, which is still popular in some cases.

In this section, we discussed the various cloud delivery models, and we learned that applications on the cloud do not necessarily have to run on someone else's computer. Rather, we can decide where we want to deliver our software, be it to our private, public, or hybrid cloud. Now that we have learned where the cloud could be, we will shed light on the question of what the cloud is and how it can be characterized in the next section.

Cloud characteristics

Today, many services that were referred to as **hosted services** a few years ago are now considered cloud services. While this is fine in many cases, some definitions help us understand cloud services and how they can be delivered. One of the most well-known specifications of these characteristics is the NIST definition of cloud computing (`https://nvlpubs.nist.gov/nistpubs/legacy/sp/nistspecialpublication800-145.pdf`), which helps us classify services but also characterizes cloud computing. Throughout this book, we will primarily refer to the definitions provided there.

In hosted services, we often had to write an email or call our service provider when we needed new services. The same applies to internal infrastructure. In the best case, the infrastructure was virtualized, and resources were available. In the worse case, infrastructure had to be ordered, and it took some time to order and install the hardware. These are things that are not imaginable nowadays. When a new runner in our CI system is needed, we assume that enough hardware is available to fulfill our needs. Therefore, we can assume that there is at least a self-service portal that enables us to create resources. In the best case, some APIs could be used to provision our infrastructure. When using a cloud service, we want it to be accessible. In our CI system, we might use external dependencies to build our software, but we also need to access internal resources over a private link. Therefore, we assume that there is enough (and, in the best case, *guaranteed*) network bandwidth when we subscribe to cloud services.

Last but not least, we only want to pay for the things we consume. As each of our runners might only run for a few minutes, we only want to pay for the time we consumed instead of paying for a whole month. As a prerequisite, we assume that such services are metered and that we know what we are paying for.

In short, and to summarize, cloud computing has the following characteristics (according to NIST 800-145):

Characteristic	Description	Example
On-demand self-service	A customer can request a service without human interaction	Trigger the provisioning of a virtual instance via an API
Broad network access	A customer can access the service over the network	After the virtual instance has been provisioned, it can be accessed via SSH or a cloud console
Resource pooling	The resources used by the services are pooled to serve multiple (internal and external) customers, and the customer has no control over where the service instance is hosted	The requested instance is hosted somewhere in a data center location, but the customer has no control over who shares the infrastructure and whether it will be started on the same virtualization host when it gets restarted
Rapid elasticity	A customer can assume that there is (almost) unlimited infrastructure available to scale on demand	An application server is high on load during peak times, and the customer adds additional virtual instances to handle the load
Measured services	A customer can find out how long resources were used for and which capacity has been used	When running a virtual machine, the customer gets billed for the time it has been running

Table 2.2 – Characteristics of cloud computing (according to NIST 800-145)

As we will see later in this chapter, these characteristics are important when we are designing our continuous delivery system for cloud-based applications. But before we get there, it's important to know which kinds of services we will deliver in the cloud.

Going further, let's discuss the benefits of cloud computing and how its main characteristics enable organizations to accomplish business outcomes.

In the subsequent table, we attempt to connect the dots:

Characteristics of the Cloud	Business Outcome	Service Offering from Cloud Service Providers
On-demand, self-service	% improvement in response time of provisioning the requested service	Cloud service providers are constantly working toward improving the average response time for users and services co-located within a region and also cross-pollinating the user base across regions as required in many cases. There are provisioning tools for various types of cloud deployments, including multi-cloud deployments.
Broad network access	% improvement in reliability by the ability to access a variety of services hosted on the cloud from different service providers and cloud-based vendor applications	Cloud providers have added different reliability tiers for different kinds of services to improve the reliability that's needed by the consumer.
Resource pooling	% optimization of cost for available capacity by enhancing utilization over a certain period	Cloud providers offer optimized resource pooling features to balance the cost by tracking utilization, capacity need, and varying demand.
Rapid elasticity	% gains from **Operational Expenditures (OpEx)** reduction by maintaining a balance between supply and demand	Cloud providers offer features to manage elasticity predictably by scaling the services up and down as needed by the users.
Measured services	% gains from increasing throughput by measuring consistently and taking actions to improve the overall throughput and productivity	Cloud providers have various trackers that can be customized. Through these customized metrics, the users can get valuable insights to take action to improve the overall throughput.

Table 2.3 – Mapping cloud computing benefits to cloud computing characteristics

In this section, we took a very short look into the cloud characteristics that are relevant to continuous delivery. With this, we can proceed further and dive into the cloud service models that may help us understand how services are deployed and who is responsible for which parts of them.

Cloud service models

In the previous sections, we discussed where and how our cloud services can be deployed. In this section, we will talk about who is responsible for which level of maintenance of service and how the most common service levels – **Infrastructure-as-a-Service (IaaS)**, **Platform-as-a-Service (PaaS)**, and **Software-as-a-Service (SaaS)** – differ. Furthermore, we will discuss the modalities that customers can expect when choosing a certain service model. We will also discuss some service models that are not part of the NIST specification but are also often used in current environments and, therefore, might have relevance for continuous delivery.

In traditional infrastructures, system and network administrators often mounted servers in racks, installed the operating system software on them, and ensured that there was enough storage and a proper network connection to provide a good service where applications could be deployed. These are the services we find in typical IaaS offerings. The cloud provider sets up an infrastructure that can have workloads running on it, obviously in a more or less isolated way. When provisioning instances in a cloud environment, we can assume that there is some network isolation between our and the other customers of this cloud service provider. Therefore, we can build configurations where our hosts can communicate with each other exclusively. For continuous delivery of infrastructure components, such as network, storage, and compute, we want to avoid consistency and integrity issues. Therefore, a practice called **Infrastructure as Code (IaC)** is adopted to accomplish this. Typical IaC tools include Terraform, Pulumi, and Crossplane.

Although IaaS is very cool for companies shifting their traditional workloads to the cloud, there is still much infrastructure-related work to do. To take this to the next level, PaaS is hiding away the infrastructure from developers so that they can focus on developing their applications, so long as they develop their applications in a way the platform supports. Nowadays, this is a very popular model, especially when a company runs a lot of services in its environment and needs much-shared functionality. A new discipline called **platform engineering** has arisen, which should enable self-service capabilities for application development organizations in the cloud-native era (read more here: `https://platformengineering.org/blog/what-is-platform-engineering`). Using this, platform engineers can provide delivery mechanisms as well as a common infrastructure (for example, service meshes and observability) to developers. As the mechanisms are used by many teams and very often, they tend to be very stable over time, and the developers don't have to reinvent the wheel every time they add another service. As this is one of the more regularly used approaches in the cloud-native world, we will focus on this throughout this book.

The third *traditional* delivery model is called SaaS and aims to only deliver the *visible* parts of the software to the customer. For example, issue-tracking systems can also be delivered *as-a-Service*. Other examples of this are services that might be used for CI/CD, such as cost analysis of IaC tools and security scanners. When consuming such services, we don't have to take care of the underlying infrastructure and the maintenance of the software. We can use this service through its web interface or APIs; therefore, the administrative effort is minimal.

Although it might sound obvious at this point to use SaaS whenever possible since the cost factor is feasible, there might be other things to consider when choosing a service model. It might sound very feasible to use managed platform services (for example, Kubernetes) whenever possible so that the **cloud service customer** (**CSC**) doesn't have to manage the underlying infrastructure and care about updates. Nevertheless, the CSC would have the requirement to be able to run on multiple cloud providers (**multicloud**), and therefore standardizing the behavior of the platform across these multiple providers might get challenging. Due to this, there is always an interrelationship between the operational effort we put into a cloud service, the direct value the service brings to us, and the flexibility in operations and the design we have when using this, as depicted in *Figure 2.1*:

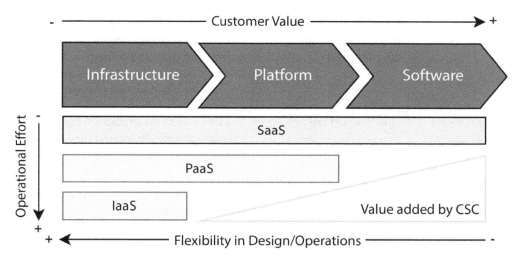

Figure 2.1 – Value versus flexibility

Over time, additional service models were developed that are not part of the NIST specification but are relevant to our delivery journey:

- **Function-as-a-Service (FaaS)**: Instead of a service on a platform, the artifact that is delivered is a simple function. Therefore, the platform on which the function will run is provided by the cloud service provider, and the function gets scheduled on this platform. A typical example could be a function that sends messages, though small batch functions based on events can also be handled with FaaS.

- **Container-as-a-Service (CaaS)**: While hosting full container platforms such as Kubernetes is often considered SaaS, CaaS allows you to schedule a single container; the provider takes care of the underlying infrastructure.

Now, it's important to map the characteristics of CD that were presented in *Chapter 1* and connect them to the cloud service model to build a connection. We have briefly touched base on this aspect in the following table:

Characteristics of CD	Description	Connection with Cloud Service Models
Speed	How frequently can we release?	Cloud service models facilitate expediting the provisioning of required infrastructure, platforms, and applications for continuous delivery
Security	How trustworthy are we?	Readily available security tools and applications can be integrated into the pipeline on-demand as SaaS to enhance security and eventually establish trust
Agility in software development	How fast can we respond to the changing needs of the customer?	Many planning tools for portfolio and product backlog planning, orchestrating sprints, and easily integrating the work items into the overall delivery process are available through the cloud service providers
Simplifying software life cycle management	How resilient are we?	The life cycle of the available services is managed seamlessly without the need to change windows or have downtime in most cases
Collaboration with software operations	How effectively do we communicate and share information between teams?	Distributed development teams are supported through various tools and integrated workflows, increasing developer productivity

Table 2.4 – Mapping CD characteristics with cloud service models

Since the previous sections were a bit more theoretical and we learned lots of basics around the cloud, in the next section, we will find out how this matters concerning continuous delivery.

Why is continuous delivery different in the cloud?

With all of the findings in the previous few sections, it might be obvious that the options for how to deliver software are endless, and there are many valid architectural approaches to design it, much more than we had in the pre-cloud era when we had virtual or physical servers, where we delivered software through remote runners or other mechanisms.

Although the infrastructure was not a primary concern of continuous delivery, it became more and more important in the past few years. Traditional infrastructure was often very static, but with the self-service capabilities and the *endless* resource pool of the cloud, we have the option to spin up new infrastructure, deploy the new version of our application, test it, and shift the traffic to the new version of the application gradually before removing the old version of the application. But to take advantage of these capabilities, cultural and technical aspects must be considered, all of which will be covered later in this book.

Furthermore, we knew that all of our services were running in our own (or external) data center, and network connections were very stable throughout the time (although as I learned in one of my first courses about distributed systems, you should never assume that a network connection is stable). When running in the cloud, there is not one target system to which we are deploying. Systems might be located in our own data center but might also run on the cloud provider infrastructure of one or many cloud providers all over the world.

Another thing we should consider is that we might deliver different types of services we deploy, and we have to take into account whether we are running on shared infrastructure (platform) or whether the infrastructure is dedicated to our service. Therefore, we could use a PaaS for our database but deploy our applications through virtual instances or on a shared Kubernetes cluster. Last but not least, simple things could be running as functions. All of these services might have different interfaces or APIs so that they can be deployed and configured, and they might face issues we have to deal with.

If you are looking for a *one-size-fits-all* solution or a recipe to build a continuous delivery solution for all of these cases, we have to disappoint you. How such a solution might look will vary based on your requirements and the nature of the applications you are deploying. The good news is that there are great building blocks and tools out there that will help you achieve all of this.

Opportunities and risks of practicing cloud-based continuous delivery

Let's start with a simple visual of the building blocks of continuous delivery in the cloud. Referring to the building block of continuous delivery (*Figure 1.1* in *Chapter 1*), we can take a simple approach to visualize the building block being hosted in the cloud:

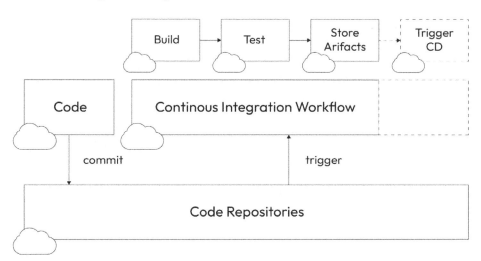

Figure 2.2 – Continuous delivery in the cloud

Now, it might look like it is a matter of location where these building blocks are hosted; however, it is more than that. Cloud-based continuous delivery can be a catalyst for multi-dimensional transformation, which comes with its own set of opportunities and challenges.

We can assume that these components are hosted in the cloud with cloud computing characteristics. Making this a little more complex, some organizations are in the middle of the transformation, which means that only some of these building blocks are hosted in the cloud, whereas for others, it can be a mix of cloud providers.

Opportunities for continuous delivery in the cloud

In *Chapter 1*, we discussed the continuous delivery ecosystem and its three components. Needless to say, CI/CD acts as a catalyst for transforming software delivery, and there is an increasing demand for new tools and applications to support CI/CD. Continuous delivery practitioners and cloud providers are constantly evolving in new ways to co-create these tools and applications through cloud-based **marketplace extensions**, where practitioners can host new tools and integrate them into the ecosystem.

Another important evolution is **pre-built CI/CD pipelines**. These cloud-hosted pre-built CI/CD pipelines are easy-to-use pipelines and remove unnecessary provisioning and configuration steps. In some cases, these pipelines can be used with simple **graphical user interface** (**GUI**)-based tools through which they can be deployed in minutes.

Onboarding to CI/CD in the cloud for organizations that are using traditional CI/CD methods can be a daunting task. With time, practitioners, along with cloud providers, have made it easier to take on this journey incrementally. Organizations can opt to go all-in with the cloud or deploy CD in the cloud in a stepwise manner, maximizing the benefits of both worlds.

Some of the functions, such as service-level management, vendor management, license management, and compliance management, can be simplified by integrating the process of onboarding continuous delivery in the cloud, where the cloud provider can act as a managed services provider.

Challenges of continuous delivery in the cloud

It might seem that continuous delivery in the cloud is becoming an obvious option for practitioners; however, it comes with its own set of unique challenges, including *reference architecture alignment and standardization of the continuous delivery framework, interoperability, and vendor neutrality*.

In the inception stages, the cost and effort to migrate continuous delivery to the cloud can take a substantial amount of time and effort for organizations that are using traditional methods. Another aspect is that even though it looks like onboarding CD capabilities to the cloud can shorten the lead time for developing and deploying new tools, applications, and features, it can also act as a limitation, where a dependency is introduced by architectural constraints and vendor lock-in.

It is worth mentioning that security, new risks, and vulnerabilities might be introduced if there is a limited understanding of cloud technologies and connected evolutions. Careful assessment and continuous improvements would be needed to secure the pipeline effectively, which can also increase the cost of end-to-end deployments.

Open source management solutions are evolving; there has been a shift to a more pragmatic approach of securing the software supply chain for open source tools and applications in particular. It is becoming evident that organizations need support from cloud-native communities to adopt continuous delivery in the cloud with open source.

Finally, new skills will be required to evolve continuous delivery. It can have a substantial impact on the operating model of the organization, and upskilling programs must be invested in.

Summary

In this chapter, we learned a lot about cloud computing itself, its characteristics, and how they might affect our continuous delivery strategy. First, we took a closer look at the deployment models – private, public, and hybrid cloud – and found that they affect our continuous delivery strategy since our software might also be spread across multiple locations.

After that, we discussed the main characteristics of the cloud and some of the differences between the cloud and typical hosted services. We learned that self-service, elasticity, and resource pooling can help us with continuous delivery by adding infrastructure on demand.

To round up the cloud basics, we mentioned the various service models available (IaaS, PaaS, and SaaS) and elaborated on their main characteristics. We also found out that we might not deliver an application in a single service model.

After, we discussed why continuous delivery is different in the cloud and what opportunities, as well as risks, could be present when practicing continuous delivery in the cloud.

The next chapter covers the prerequisites for onboarding to continuous delivery in the cloud from an organizational perspective. You will be provided with an introduction to various aspects of organizational readiness and learn how to lay out successful strategies for continuous delivery in the cloud.

3
Creating a Successful Strategy and Preparing for Continuous Delivery

This chapter covers in detail the prerequisites for onboarding to **continuous delivery** (**CD**) in the cloud from an organizational perspective. It introduces various aspects of organizational readiness and explains how to lay out successful strategies for CD in the cloud. This chapter guides you through the process of implementing, managing, and improving CD in the cloud. It also digs deeper into preparing for the requirements of modernization for CD with regard to the cloud, the core challenges involved, aligning and connecting the roadmap with the cloud to maximize the return of investment of CD, and lastly, competence and upskilling needs.

The chapter covers the following topics:

- Strategic goals and objectives of modernizing CD in the cloud
- Connecting business, transformational, and operational strategies
- Outlining organizational prerequisites for CD in the cloud
- Upskilling goals for the organization
- Modernizing for CD in the cloud

Strategic goals and objectives of modernizing CD in the cloud

Investment in CD is a critical prerequisite for business modernization efforts. It is expected that by adopting CD, organizations will significantly enhance their software delivery productivity, which in turn enables the delivery of new products and services. CD could thus be considered a force multiplier. However, many organizations find that they cannot exploit the full potential of CD at scale. There are many possible reasons for this strategic gap between expectation and reality. Let's start with understanding why this happens and how investing time in setting the organization's strategic goals and objectives can help close the gap:

- **Competing priorities**: Organizations can have siloed initiatives where many different segments of the organization each try to steer the cloud onboarding, but often in a way that is disconnected from the development and software modernization initiatives. Business leaders are narrowly focused and there are limited synergies, which constrains the overall outlook for CD in the cloud.

- **Speed is a complicated factor too**: Many organizations have rushed into DevOps thinking, lacking advanced planning, roadmaps, and retrospectives. Often, organizations are stuck between two states, trying to balance their legacy setup with modernization efforts. This limits their time available to innovate and think pragmatically to improve the CD and cloud-onboarding initiatives.

- **Limited evolution during execution**: Many organizations focus on the cost and speed of software delivery but overlook other important factors. This can lead to tactical thinking that does not align with business objectives. Additionally, a lack of evolution during the onboarding of CD can lead to a regressive posture.

- **Technology-first approach**: The trade-off of adopting emerging technology to create more opportunities and associated high-level risk, including skill gap, security concerns, and so on, is always a challenge. Organizations need to shift their approach to a more people-centric view where in most cases investment is needed to upskill, reskill, and fill the resource gap before moving ahead.

- **Lack of engineering culture**: Often, organizations struggle to create the right foundation on which to develop new products and services. CD in the cloud is a combination of emerging technologies and requires a software engineering culture more than anything to thrive in the competitive ecosystem and maximize the benefits.

Having understood some of the key pitfalls of CD and cloud adoption, it is time to discuss how organizations can avoid these traps. It all starts with spending time on planning, engaging with business stakeholders to create shared goals and objectives for the long term. A **strategic goal** is a long-term business outcome. **Objectives** are measurable, specific actions connected to the strategic goal. Creating an overall map of strategic goals and objectives can be a daunting task.

In general, it is important to approach strategy with a common vocabulary as it is an organization-level process that sets the direction for the business. A good strategic plan is based on simple yet connected characteristics, and it should evolve with time. Evolving our strategy requires some sort of acceptance that at some point we may have to learn from mistakes, and discontinue the old version in favor of the new one.

Let's start with a simple illustration of how to draw up a strategic plan with interconnected goals and objectives:

Characteristic	Description	Example
Purpose	Articulate customer values aligned with your organizational values.	Accelerate the organization's ability to digitally transform
Long-term and forward leaning goals	Related business outcomes that are aligned with the purpose and provide your organization's strategic advantage.	Enhance performance of software delivery in the organization
Actionable objectives	How business, software engineering, and other related teams orchestrate value. How can you rapidly iterate and evolve to create that incremental value?	Quick release of software features and rapid response to any failures
Measurable	How progress and success will be quantified and measured.	Release frequency of new features and bug fixes

Table 3.1 – Simple illustration of a strategic plan

The preceding table is very basic. In the real world, there can be many different goals and objectives that people might want to achieve, depending on their perspective.

It is important that goals and objectives should be prioritized through tangible means. The prioritization will depend not only on the end state but also on the current situation—which goals make more sense given the existing reality.

That is the reason that these goals and objectives can differ for different organizations even if the end state is the same. Organizations often use structured approaches to prioritize and measure goals and objectives. In the subsection ahead, we highlight some key frameworks used for this purpose.

Key frameworks for strategizing CD in the cloud

There are many frameworks used by organizations to transform into their desired end state by steering the right goals and objectives. We will briefly describe some of these methodologies here:

- **Objectives and Key Results (OKRs)**: This is a simple framework to set, track, and measure goals repetitively. The concept was created by Andy Grove, but popularized by John Doerr (read more here: https://hbr.org/2018/05/how-vc-john-doerr-sets-and-achieves-goals). It is often used to set stretch goals and threshold success. **Stretch goals** are the building blocks of radical transformations and extraordinary achievements. Adoption of OKRs and specific approaches can vary across different organizations, with companies such as Google having adopted OKRs successfully. It starts with setting up three to five objectives and up to three key results at the organizational level, and then teams commit to the organizational-level objectives and set team-level OKRs aligned to the main OKRs.

- **Improvement Kata**: This framework emphasizes practicing routine. It is often used for complex systems with dynamic conditions. In this method, teams closest to the problem can make their own decisions and maneuver effectively. With Improvement Kata, the teams work iteratively toward the target state on the way to challenge learning.

- **Balance Scorecard (BSC)**: This is one of the most popular frameworks that primarily includes four perspectives: Financial, Customer, Internal Process, and Learning and Growth. Although the overall principle remains the same, that is, primarily visualizing the strategic goals and converting them into tangible objectives through which performance is measured, this method helps to break down the goals into measures, measures into projects, and projects into action items.

- **Business Model Canvas (BMC)**: This is a single-page template to outline goals and objectives for the business. BMC has nine building blocks. On the left side of the canvas it has information about the business itself, such as partners, resources, key activities, and cost structure. The middle part has the value proposition, and the right side focuses on customers and market information. It is a simple tool often used by entrepreneurs and start-ups.

Simple illustrations of frameworks for CD in cloud initiatives

Now let's take one step further in our thinking about CD in the cloud. It is always up to the organization to set the goals and drive business outcomes. There are no universal standards for putting together the goals and objectives for CD in the cloud. However, here we attempt to apply some of the frameworks we've just seen as an example:

- **Continuous Delivery in the Cloud Canvas**: This canvas is a simple illustration of BMC (more precisely, the Lean Canvas) to provide a perspective on how it can help save time and structure the strategic planning process. The following canvas illustrates the application of BMC to CD in the cloud:

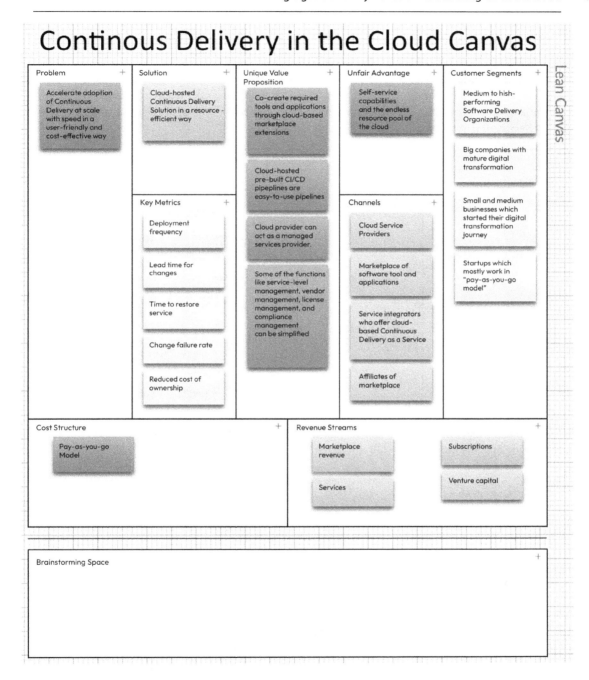

Continous Delivery in the Cloud Canvas

Problem

Accelerate adoption of Continuous Delivery at scale with speed in a user-friendly and cost-effective way

Solution

Cloud-hosted Continuous Delivery Solution in a resource - efficient way

Key Metrics

Deployment frequency

Lead time for changes

Time to restore service

Change failure rate

Reduced cost of ownership

Unique Value Proposition

Co-create required tools and applications through cloud-based marketplace extensions

Cloud-hosted pre-built CI/CD pipeplines are easy-to-use pipelines

Cloud provider can act as a managed services provider.

Some of the functions like service-level management, vendor management, license management, and compliance management can be simplified

Unfair Advantage

Self-service capabilities and the endless resource pool of the cloud

Channels

Cloud Service Providers

Marketplace of software tool and applications

Service integrators who offer cloud-based Continuous Delivery as a Service

Affiliates of marketplace

Customer Segments

Medium to hish-performing Software Delivery Organizations

Big companies with mature digital transformation

Small and medium businesses which started their digital transformation journey

Startups which mostly work in "pay-as-you-go model"

Lean Canvas

Cost Structure

Pay-as-you-go Model

Revenue Streams

Marketplace revenue

Services

Subscriptions

Venture capital

Brainstorming Space

Figure 3.1 – Simple illustration of adoption of the Lean Canvas for CD in the cloud

The Business Model Canvas was proposed by Alexander Osterwalder based on his book, *Business Model Ontology*. The Lean Canvas, on the other hand, was proposed by Ash Maurya.

- **OKRs**: An OKRs best practice is to start from the top organizational-level objectives and then cascade down to the teams. Here is an illustrated example of a mid-sized organization that has initiated its digital transformation journey. Let's examine how CD in the cloud is connected to the organization's overall strategy through OKRs:

Level	Objective	Key results
Top level (CxO)	Accelerate the organization's ability to digitally transform	• Key result 1: Launch two new software product offerings by Q2 • Key result 2: Grow revenue by $X million • Key result 3: Reduce churn by 20% by Q1
Product management	Delight customers with software product	• Key result 1: Reduce lead time to features by 10% by Q1 • Key result 2: Improve velocity of bug fixes by 50% by Q2 • Key result 3: Reduce the cost of deployment by 40%
Software delivery organization	Accelerate the adoption of CD in the cloud	• Key result 1: Reduce deployment frequency from 8 hours to 30 min by Q1 • Key result 2: Decrease downtime to 5 min during release by Q2 • Key result 3: Increase resource utilization from 40% to 60% by Q3

Table 3.2 – Simple illustration of OKR cascading

Thinking about OKRs as strategic themes and connecting the dots to improve commitment and adoption can lead to aligned action. In the preceding example, we showed how to cascade the OKRs to different organizational units in one unified way. For the adoption of CD in the cloud, look at the strategic themes and assess whether it makes sense to go in that direction, and consider who should be responsible for getting commitment from the team. There is no hard-and-fast rulebook for OKRs and no one-size-fits-all solution.

With that, we move to our next section, connecting the business, transformational, and operational strategies for CD in the cloud.

Connecting business, transformational, and operational strategies

Next, we need to create a strategic plan that aligns the goals of business, operations, and software delivery. Cloud-native software has changed the way we release software, and agile, DevOps, and lean practices have made it possible to deliver value more quickly. This has led to a need for more complex prioritization and strategy alignment, and cloud-based infrastructure has made it easier to shift strategies at all levels.

The goal of adopting CD in the cloud is to enable an incremental value flow in a fast, reliable, and cost-efficient way. Leveraging cloud capabilities and integrating it with mature CD practice is one way to achieve this, where adoption of a cloud-hosted CD solution can support the teams to deliver working software right from the first sprint:

- **Leveraging the cloud to maximize CD**: Cloud-hosted products are not built in a monolithic way, nor are they operated like monolithic software. The adoption of cloud-native principles to design and deliver software starts with a focus on scalability, resiliency, manageability, observability, and automation—the five things that make up cloud-native applications. This aspect will be elaborated further in *Chapter 5*.

- **Cloud providers can act as managed service providers**: Embracing **software-as-a-service (SaaS)** capabilities, most modern-day products support a subscription-based model of software through the public cloud marketplace. Cloud providers and third-party cloud solution integrators act as partners in the journey. Often, this requires building long-term partnerships with different vendor organizations, so careful assessment and integration is necessary to be successful.

- **Some functions such as service-level management, vendor management, license management, and compliance management can be simplified**: Reducing the in-house effort for common foundational activities allows teams to focus on the core aspects of product development and operations. It is important that CD in the cloud is viewed in a unified way and the contract agreements with various vendors should be reviewed and negotiated using a standard process. The stakeholders should build consensus about the provisions and commitments around service-level agreements, license usage, and compliance management.

- **Co-create the required tools and applications through cloud-based marketplace extensions**: Co-creating the required tools and applications through the cloud-based marketplace extension is a growing trend. This does not come without challenges, however. First, it is an emerging system; second, while the contribution to and ownership of applications when put together creates limitless possibilities, it is also complicated in the real world. There are various types of co-creation models including community-based contributions, crowdsourcing, and expert co-creation, among others. If your organization is willing to explore co-creation, it is important to evaluate the pros and cons of the various possibilities.

In this section, we touched on connecting the dots between various aspects of CD and the business outcomes. As it can be challenging to build synergies between business and operational strategies, it is recommended to collaborate in the early stages of the modernization of CD to keep it together. In the next section, we will discuss organizational culture and some key prerequisites for transforming an organization's continuous delivery posture toward the cloud.

Outlining organizational prerequisites for CD in the cloud

CD in the cloud can be a game changer; however, the journey of adoption and maturation can be painful, as *old habits often persist*. From the outside, it may appear that this process can be based on plug-and-play tools and applications from the cloud, but there is more to it. In this section, we highlight the prerequisites for adopting advanced CD practices in the cloud with specific guidance practices, approaches, and key considerations for industry practitioners.

Co-existence of Agile, DevOps, and cloud technologies and practices

If you are reading this book, Agile, DevOps, and lean practices might not be new to you. These practices are among the first prerequisites to set up and evolve CD in the cloud. More important is to put these practices together in a software delivery context:

Practice	Description	Key Considerations
Agile	A set of values and principles for better developing software	Adoption of Sprint-based planning and execution. Limiting work in progress and reducing the batch size to release working software incrementally.
DevOps	Encourage cross-functional collaboration	DevOps practices include **continuous integration** (CI) and CD and deployment, release strategies, **Infrastructure as Code (IaC)**, and observability.
Cloud	Leverage cloud-based infrastructure	Embrace a SaaS approach. Cloud-based infrastructure largely relies on automation for provisioning and configuration.
Open source	Increased open source security awareness and supply chain attack resistance	Use open source for fast access to innovative technologies, reduced vendor lock-in, and reduction of up-front costs.

Table 3.3 – Co-existence of Agile, DevOps, and the cloud

Let's move to the next one.

Focus on engineering excellence

An organization's ability to embrace engineering practices is critical to modernize software delivery, from an understanding of prioritization of business outcomes over product features to putting together practices, tools, and applications to facilitate incremental value:

- **Release orchestration strategies**: There are many engineering practices that enable an organization to release on demand. Putting together a release strategy for product features and assessing a variety of engineering options such as dark launches, feature toggles, deployment automation, selective deployment, self-service, version control, and Blue-Green deployments is a must.

- **Automation in the cloud**: Reduce manual toil on provisioning, configuration, and other redundant tasks to free up teams' bandwidth as a first step. For more mature organizations, automation also involves understanding the flow and removing bottlenecks. This requires effort to modernize architecture and facilitate the automation of high-value tasks. Another aspect is self-service and enabling it for timeliness and on-demand availability of resources.

- **Observability**: Observability provides insights in the forms of metrics, traces, and logs—the three pillars. For cloud-native applications, observability is one of the most effective practices to understand what's happening across all of your environments and technologies, so you can detect and resolve issues to keep your systems efficient and reliable. You can read more here: `https://www.cncf.io/blog/2022/05/31/what-is-continuous-profiling/`.

- **Site reliability and resiliency with the cloud**: Resiliency is the ability to adjust the functioning of a system during both expected and unexpected conditions. One of the principles here is to spend 50% of your time doing ops and 50% on dev tasks, bringing the team closer for cross-functional engagements. You can read more here: `https://cd.foundation/blog/2023/03/01/evolution-of-ci-cd-with-sre-a-future-perspective/`.

- **Future pipelines in the cloud**: CD and connected practices are evolving as a result of cloud technologies, tight collaboration to set the interoperability standards, and the use of observability to assess the future trends as event-driven orchestration, managed by policies and GitOps. You can read more here: `https://cd.foundation/blog/community/2022/12/06/pipelines-pipelines-pipelines/`.

Ability to collaborate with the legacy, monolithic world

It is evident that most organizations are in a hybrid state with pockets of legacy and monolithic environments existing alongside modern ecosystems. Resolving this situation requires strategic alignment and upfront planning of how the legacy will be transformed within due course. Determine whether the current technical debt is sustainable or whether specific investments are needed to address it.

Operating models embrace third-party service providers

Onboarding to cloud-hosted pipelines and platforms calls for a significant rethinking of governance structure, people strategy, and how relations with third-party service providers are handled. Distributed architecture comes with distributed teams, split across multiple time zones. Another aspect is outsourcing some of the capabilities, such as automation, pipeline design and delivery, specific knowledge about cloud services, and so on, to get early access to the talent pool with limited risk. All these changes to the operating model can become overwhelming for leadership and careful due diligence is needed in more than one dimension.

Communication and collaboration

With a multi-stakeholder environment comes the challenge of effectively communicating and collaborating. Creating a long-term partnership with cloud providers and keeping management overheads below 10% by contractually outsourcing the most common functions to the provider is a daunting task as well. Communication plays a vital role in a self-organizing context.

We conclude this section now, noting that it is extremely important not to overlook the organizational prerequisites to avoid burnout and frustration within the teams. In the next section, we will highlight another aspect of the modernization journey: right-sizing the team and how we can achieve it through thought leadership around upskilling and creating cohesive team structures.

Upskilling goals for the organization

This section provides insight into right-sizing a CD team that is decentralized, distributed, and self-organizing.

Organizing for CD in the cloud

Before we dig deeper into the roles and structure of the team, and the upskilling requirements of team members, let's look at some of the anti-patterns in the ecosystem when it comes to team structures:

- **Organizing around skills, such as a technology and architecture team, security team, and operations team**: It is highly likely that there will be siloed initiatives, tools, and processes with long lead times for release and decision paralysis with many stakeholders driving disconnected outcomes.
- **Organizing around projects**: It is always difficult to resource multiple competing projects with the right skillsets. Organizing around projects eliminates the siloed teams' problems but creates technical debt, thus focusing on the deliverables but often leaving operations behind.

Now let's look at some team structures more aligned with modern software needs. For ramping up and looking at the upskilling goals, it is important to understand the target end state of the operating model and how that could influence your upskilling choices:

- **Organizing the teams around business or product lines**: This setup is focused on roadmaps, product planning, and having a mature understanding of the business outcome. Often infrastructure and common foundational capabilities are either outsourced or considered low priority. There is a risk that these capabilities become a showstopper due to the lack of alignment and investments.

- **Cross-functional full stack teams**: In this type of team, the organization takes a full stack approach. The team has a mix of interdisciplinary skills to organize itself and accomplish the business outcome. This type of structure supports agile, DevOps, and lean thinking and consists of the application, infrastructure, and tools knowledge to adopt CI/CD in the cloud. It can become overwhelming once the initiatives start to grow and careful alignment is needed with the strategic goals and business outcome on a regular basis.

In the following figure, we illustrate the different components of an operating model toolkit for modernizing CD:

Operating Model Toolkit

Strategic goals and objectives for modernizing Continuous Delivery

Purposeful	Long-term and forward leaning	Actionable objective	Measurable

Operating model and organization components

Practices	Structure and teams	Engineering excellence	Metrics
- DevOps - Agile - Lean - Cloud	- Organizing the teams around product lines - Cross-functional full stack - Embracing third party service providers - Governance and visualization	- Release orchestration - Observability - Automation - Site reliability and resilience engineering - Pipeline in cloud	- value - flow - scalability - resilience - manageability

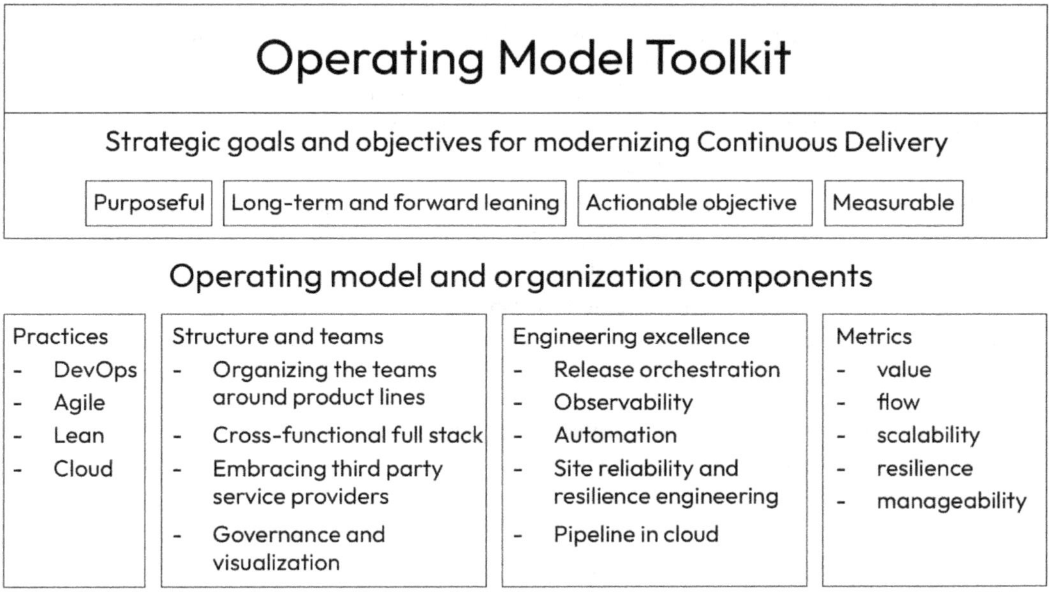

Figure 3.2 – Illustration of an operating model toolkit for modernizing CD

We will now discuss key skills for individuals and teams.

Key skills to improve CD posture in the cloud

In addition to the key skills for individuals and teams, we will also highlight some aspects of governance and the skills that can help you build a robust posture for CD:

- **Embrace CD as a core competency**: Thriving in the new ecosystem requires a mindset shift at all levels. Embracing the CD principles is a foundational competency for organizations that wish to reach the next level of maturity.

- **CI/CD as a portfolio capability in the cloud**: Start treating CI/CD as a product. The roles and competencies needed for a product team are not any different from a regular product: a product manager, product owner(s), and CI/CD practitioners and engineers:

  - A **product manager for CD** is responsible for the roadmap and general incremental features for application stakeholders.

  - **Product owner(s) for CD** are responsible for the resilience and interoperability of the pipeline.

  - **CD practitioners** are responsible for seamlessly integrating features from different service providers and tools in a unified way. They are also responsible for the reference architecture and shared pipelines.

  - **CD engineers** are primarily responsible for the evolution of CI/CD practices.

- **Application engineering**: Application engineering comprises full stack engineers who can use the services offered for incremental deliveries. The application engineering team creates workflows, environments, and integrations with CD pipelines as needed.

- **CD governance and visualization**: One of the most important aspects of CD pipeline governance includes policy management, audit trail capability, integration of full stack observability, security, and supply chain management. Translating data into insights and visualizing the health of CD capability including cost management is also important. Considering governance around costs can help develop innovative ways to move toward cost excellence when onboarding CD in the cloud.

In this section, we discussed the upskilling aspects connected to CD transformation in the cloud. In the next section, we will briefly introduce the implementation, operations, and CI aspects of a cloud transformation journey.

Modernizing for CD in the cloud

One of the most fulfilling parts of a transformation journey is to actually perform the implementation and subsequent management of the change. It is easy to fall back and revert to old ways if insufficient attention is paid to streamlining the implementation and management of the change being introduced through modernization. In this section, we will highlight briefly some aspects of advancements of CD in the cloud as regards implementation.

Implementation

With the advancement of CD practices, tools, and processes, we have an array of applications and tools to choose from. This presents a set of benefits including not limited to the following:

- The ability to apply CI and CD to software development through a set of abstractions without the need to build them from scratch

- Allowing operational oversight of releases and versions

- Treating all processes as code within the pipeline

- Subscription-based adoption of capabilities on demand

On the other side, using multiple tools with different capabilities can also lead to challenges. For example, redundant tools for serving the same purpose, often monitoring tools that are applied to subset of the CD pipeline, is one area where organizations implement different tools serving the same purpose.

While implementing and managing the modernization of CI/CD, it is important to think about the tools and processes that must be loosely coupled and vendor agnostic.

Managing CD

As organizations adopt CD at scale, it is critical that standardized guidelines and policies are developed and maintained to implement CD practices. As the adoption of CD is not standardized across organizations, tool vendors, and cloud providers, it is advisable to work on the following key areas to manage and scale CD at the enterprise level:

- **Shared responsibility model**: Defining the line between the responsibilities of your organization and those of your providers is imperative for reducing the risk specifically for cloud-hosted components.

- **Avoidance of vendor lock-in**: Vendor lock-in is a situation where an organization using a product or service cannot easily transition. Proper assessment during vendor selection and careful build-versus-buy decisions could help to reduce time and effort when managing CD at scale.

- **Unified observability to improve resilience**: Deep integration of observability for CD can support advanced automation and event-driven orchestration of pipelines.

- **Cost management**: CD also has a life cycle and cost implications. Taking a product-oriented mindset, evaluating the incremental features, and creating a business case for every incremental drop helps manage the cost of CD. Business analytics of operational costs helps optimize the cost of running the CD stack via increased efficiency.

Management metrics

One of the critical success factors for cloud-based CD is measuring what matters and course-correcting when required. One of the popular ways to measure CD performance is using metrics developed by DORA's Accelerate State of DevOps research program. The program has validated four elements that measure software delivery velocity and a fifth one for stability:

- Deployment frequency
- Lead time for changes
- Mean time to restore
- Change failure rate
- Reliability

Many cloud-based CD tools and platforms offer out-of-the-box DORA metrics visibility for teams to measure their current state, deliver visibility across the value chain, streamline business objectives, and promote a collaborative culture.

You can read more at `https://dora.dev/publications/`.

Trending cloud-based CD tools

Let us look at a few tools:

- **Harness**: This tool (or platform) provides continuous integration, delivery, and deployment capabilities along with cloud cost management. It integrates with resources including cloud platforms and repositories and tools such as Jenkins, Jira, ServiceNow, APMs, logging aggregators, and Slack.

- **GitLab**: This tool (or platform) provides a zero-touch CI/CD capability in one integrated platform. Some benefits of using a single unified platform for the entire software life cycle are better visibility, simplicity of use, and having one single source of truth. GitLab has a SaaS offering that can integrate with a number of tools and applications hosted on the cloud.

- **Spinnaker**: Spinnaker is an open source, multi-cloud CD platform that combines a powerful and flexible pipeline management system with integrations with the major cloud providers.

- **GitHub Actions**: GitHub Actions is a SaaS product built right into GitHub with no other software needed. GitHub Actions subscribes to a lot of paradigms made popular by other tools and pipelines such as YAML, building in containers, and so on.

- **Azure DevOps**: This is a Microsoft product that provides version control, reporting, requirements management, project management, automated builds, testing, and releasing management capabilities hosted on the cloud. It has some predefined community-built pipeline jobs that you can plug and play.

- **AWS CodePipeline**: This is a fully managed CI/CD service from Amazon Web Services that helps you automate your release pipelines for fast and reliable application and infrastructure updates.
- **Google Cloud Build**: This is Google's own CI/CD platform.

With this, we conclude the final section of this chapter. CD practices, tools, and even the architectural aspects are evolving and it is beneficial to use these to create a progressive culture of CI. We will discuss in more detail some of these technical aspects in the upcoming chapters to provide more insights into the advancements and how to keep up with the evolution.

Summary

With the understanding of CD's main pitfalls and challenges of the past, as well as the key next steps for managing an advanced CD posture in the cloud, this chapter provided insights to help you plan, implement, manage, and measure the maturity of your CD strategy. The recommendations for appropriate adoption of future practices, tools, and skills will help save teams time. Establishing foundational capabilities can further support scaling cloud-hosted CD in an interoperable way.

In the next chapter, we will dig deeper into setting up and scaling a CD project on the cloud.

4

Setting Up and Scaling Continuous Delivery in the Cloud

This chapter provides readers with a detailed view of setting up **continuous delivery (CD)** in the cloud to maximize the overall **return on investment (ROI)** on software development, infrastructure, and management. This chapter provides readers with an overview of scaling up CD in the cloud for enterprises. It includes organizational characteristics, requirements for scaling CI/CD (where **CI** stands for **continuous integration**), and ownership of CI/CD with roles and responsibilities. The reader will gain an understanding of various key aspects of cloud onboarding of CD at a high level. The reader will be introduced to the maturity curve of CD, and the chapter will help them to be successful in the evolution toward cloud-based CD.

We will cover the following topics:

- Choosing the right cloud environment, model, and tools

- Understanding and owning the culture for CD in the cloud

- Shifting left—cost, risk, and security management

- CD practice and its evolution

- Key requirements for scaling CI/CD

Choosing the right cloud environment, model, and tools

Today's market for cloud providers is quite dynamic and evolving. Choosing the right cloud environment is not a subjective topic; however, there are resources available to help you make the right decision for your context. Let's first discuss the process of selection of the environment, model, and tool.

We intend to cover the steps including preparation, strategic analysis, initial selection criteria, and—lastly—how to adopt and manage the evolving CD practice in the cloud. This section provides high-level support for decision-makers to follow a pragmatic approach.

Preparation

Organizations have to prepare individuals, teams, and the organization itself for embracing change. Consider the big picture, top goals, and objectives for the organization; also, perform stakeholder interviews, user research, and—lastly—technology and business research. There are frameworks, techniques, and methods that help individuals and organizations to prepare for choosing the right tool, model, and environment from the inception stages. We present a simple explanation of how you can initiate the process of selection of cloud environments in a structured way in the following table:

Preparation Process	Questions	Key Objective
Big-picture view—goals	What is the primary business outcome of moving to the cloud? Does modernizing CD in the cloud support the primary business outcome?	Clearly articulating the strategic goal
Stakeholder interview	Which **business units** (**BUs**) and products are likely to benefit from cloud onboarding? What are the key quantitative benefits of adoption of a particular cloud environment, tool, or model? Can your products remain competitive without using the cloud? Does it help refine the operating model to improve margins?	Exploration of stakeholder's perspective to align and fund the right initiatives
User research	How accessible and easy to use are cloud offerings? How reliable is the offering?	Deliver the best value aligned with the big picture and accelerate adoption

Technology research	Which technology options bring sustainable results in the long term?	Identify quick wins and stay ahead of the curve
	Is it feasible to buy when we can build it ourselves?	
	Does the cloud provider's product and services roadmap align with our needs?	
	How comprehensively does the cloud provider address security, governance, and compliance requirements?	
Business research	Are we compliant with regulatory onboarding to a particular provider?	Sustain business outcome
	what is the financial health of the cloud provider like for **business continuity (BC)** in the long term?	
	How complex are contracts, agreements, and third-party service dependencies? How efficiently is contract exit supported?	
	Does the offering support a business model that improves cash flow?	

Table 4.1 – Simple illustration process of preparing for selecting a cloud environment

Strategic analysis

Once the organization has done the initial research and collaborated with different stakeholders, it's then time to start the process to assess various options. Various methods can be used to perform this assessment, such as doing a **SWOT analysis** or **scenario planning**. Here, we present a very simple yet effective way to perform a strategic analysis of different cloud provider options that can be used as a starting point for an assessment.

Defining the key criteria

The next step in the process is to set a criterion for the evaluation of products and services. As there is no one-size-fits-all cloud capability, we provide some guidance in this section around the criteria of assessment. Here are some key focus areas:

- **Products and service offering**: Outline the spectrum of products and services that is needed currently or in the near future for evolving practices around CI/CD—for example, tools and services required for container orchestration and infrastructure and application management tools and services. One such tool is GitOps. It is equally important to assess the infrastructure and platform services alongside off-the-shelf applications required to build the end-to-end CD pipeline.

- **Operating cost**: One of the most compelling benefits of cloud-based offerings for BUs is more efficient cash flow. There are many ways to achieve this, such as shifting from asset ownership to consumption-based service contracts. This might result in lowering the upfront **capital expenditure (CapEx)**. Resource pool optimization is another aspect contributing directly to efficiently managing the operating cost. Another aspect is a lower **total cost of ownership (TCO)** by buying plug-and-play capabilities instead of building them from scratch. Cost predictability and transparent chargeback are other aspects to look at when selecting a cloud provider.

- **Reliability**: A mature cloud provider would have a good track record for service reliability. The cloud provider should maintain a history of periodic component failures and cloud service downtimes to assess the performance and impact. Some examples of such dashboards are listed as follows:

  - **AWS**: `https://health.aws.amazon.com/health/status`

  - **Azure**: `https://status.azure.com/en-us/status?WT.mc_id=A52BDE99C`

  - **Google**: `https://status.cloud.google.com`

  There is third-party software that can provide similar functionality to assess the performance of the entire stack.

- **Security**: It is important to have a mature security posture with a deep understanding of designing, implementing, and operating cloud services securely. It is also important to understand the modern threat landscape and introduce modern features to secure the software stack from vulnerabilities. Some key areas to assess would be network and infrastructure risks, **identity and access management (IAM)**, integration of security information and event management, and—above all—a DevSecOps culture to ensure cross-functional alignment on implementation of security practices integrated from inception stages.

Apart from these criteria, long-term alignment between the organization and cloud provider roadmap, close collaboration, and healthy contractual agreements are a baseline for sustainable business outcomes. It is important that the cloud provider has a healthy financial status to execute the roadmap and customer contracts. Other key aspects that regularly evolve are the architecture, product features, and infrastructure cost model based on customer feedback. Another aspect is the support capability for services and flexibility in offering different types of support models. It is necessary to ensure that the **cloud service provider** (CSP) follows regulatory guidelines and standards and is certified. Some industry standards such as the **General Data Protection Regulation** (GDPR), **System and Organization Controls 2** (SOC 2), the **International Organization for Standardization** (ISO) *27000*, the **Health Insurance Portability and Accountability Act** (HIPAA), and the **National Institute of Standards and Technology** (NIST) can be starting points for assessing the regulatory posture. Another aspect is the adoption of interoperability standards to avoid vendor lock-in situations.

Initial selection

It is evident that choosing a foundational CSP is one of the first steps in the process of modernization of the spectrum of tools and applications required for CD. One way to make an initial selection is to associate weights to each of the criteria based on their importance; this can be baselined according to the data collected in the first step in the preparation section through user research and stakeholder interviews. As, by now, you must have understood there is no one-size-fits-all solution, careful due diligence is required to map weights to each of the criteria to facilitate the selection. In the following table, we present a simple example of weighted criteria for selection that can provide you with a basic idea of structuring your own criteria in a data-driven way:

Criteria	Weight	Guidance
Cloud-based products and service offerings by provider	High	A spectrum of products and services offered and seamlessly integrated is of prime importance if you are trying to set up cloud CD as a first step.
Operating cost	Medium	The importance of operating cost is directly proportional to scalability. Once the organization starts to scale, this criterion becomes more important.
Reliability	High	Reliability is one of the criteria that remains of high importance irrespective of the maturity and scale of onboarding.
Security	High	Security of cloud-based services is of high importance irrespective of the maturity or scale.

Table 4.2 – Simple example of weighted criteria for selection

Here are some additional criteria to consider in regulated sectors:

Criteria	Weight	Guidance
Sustainability	Medium	Once the onboarding to cloud-based delivery matures, this criterion becomes significantly important.
Certifications and standards	Medium	For highly regulated sectors, certification and standards are very important.

Table 4.3 – Extended example of weighted criteria for selection

Some more guidance is available through various research reports—for example, *Gartner Magic Quadrant* analysis uses a similar assessment technique paired with SWOT analysis.

Adopting and managing a CD practice

To grasp the wide range of benefits of CD, it is important that we continue to evolve and embrace advanced capabilities offered by different products, Let's have a look at this in more detail:

- **Products and services**: CI/CD tools work with a pipeline strategy and overtake the manual steps of getting the latest changes from the source code repository, compilation, testing, verification, and deployment to the cluster.

- **CI/CD pipelines in a Kubernetes ecosystem**: Container orchestration has undergone evolution with Kubernetes. Various tools support the management, security, and monitoring of container orchestration. Modern CI/CD pipelines treat infrastructure components as an integral part of application development and also bring the practice of managing infrastructure closer to application management practices. To manage the infrastructure components, there are pull-based and push-based pipelines. The rise of GitOps has also contributed to enabling a declarative approach to infrastructure management.

- **Next steps with infrastructure as code (IaC) for CD in the cloud**: As mentioned in the previous point, both infrastructure and application management are now evolving toward declarative approaches and providing developers with a unified way to handle updates on both applications and infrastructure. Infrastructure components are evolving to a more versioned and immutable state. To add to this, improvements in IaC components and templates with self-healing capabilities ensure the configuration and provisioning are optimized. Improving developer productivity with future-ready pipelines, mostly modular, helps build on interoperability standards and is automated through events in the CI/CD workflow. Projects such as CloudEvents and CDEvents are actively working to define interoperability through the standardization of events across the pipeline. For further reference, visit `https://github.com/cloudevents/spec/blob/main/cloudevents/primer.md` and `https://github.com/cdevents`.

- **Open source CI/CD pipeline**: There are many popular open source tools to orchestrate modern CI/CD capabilities. Some of these CI/CD tools include the following:

  • **Spinnaker**: Spinnaker is an open source, multi-cloud CD platform that combines a powerful and flexible pipeline management system with integrations to the major cloud providers. If you are looking to standardize your release processes and improve quality, Spinnaker is for you. The platform implements in-built, production-safe deployment strategies to help deliver applications to their target environments without having to write any scripts.

  • **Argo CD**: A declarative, Kubernetes-centric CD tool that fetches commits made in a centralized repository and applies them to the production cluster. It is dependent on a pull-based mechanism.

  • **Jenkins X**: Jenkins X automates and accelerates CI/CD for developers on the cloud. Open source projects that Jenkins X integrates with include Kubernetes, Tekton, Grafana, and Nexus. It is limited to Kubernetes clusters and works purely on GitHub.

- **CI/CD pipeline with cloud providers**: Cloud providers also offer CI/CD tools; some of these tools can help you create and automate entire pipelines, while others are better at specific tasks. The following table gives an overview of the cloud provider-specific tools available within the cloud providers. Several AWS, Google, and Microsoft services in this section are not explicitly available through either Google Cloud or Microsoft Azure; they are part of each vendor's broader cloud portfolio:

Capability	AWS	Azure	Google
Artifact management	AWS CodeArtifact	Azure Artifacts, GitHub Packages	Artifact Registry (preview)
CI/CD	AWS CodeBuild, AWS CodeDeploy, AWS CodePipeline, AWS CodeStar	Azure Boards, Azure DevOps, Azure Pipelines	Cloud Build, Tekton
Private repository	AWS CodeCommit, AWS Serverless Application Repository	Azure Repos	Cloud Source Repositories
Testing	AWS Device Farm, AWS Fault Injection Simulator	Visual Studio App Center, Azure Test Plans, Azure Internet Analyzer (preview)	Google Firebase Test Lab

Table 4.4 – Overview of cloud provider CI/CD tools

- **Security**: With the modernization of software architecture, more and more software products consist of commercial software components and open source coexisting, producing, sharing, and consuming a **software bill of materials (SBOM)** and implementing related standards. Modern CD product vendors are improving how SBOM is supported to improve the security posture of software applications by enhancing software supply chain transparency. The **National Telecommunications and Information Administration (NTIA) Software Transparency Working Group on Standards and Formats** is actively working along with industry experts to evolve standards to facilitate their adoption. The **Open Source Security Foundation (OpenSSF)** is another initiative that is contributing to the maturing posture of SBOM. For further reference, see `https://ntia.gov/sites/default/files/publications/sbom_formats_survey-version-2021_0.pdf`.

- **Operating cost**: Operating cost is one of the key areas for management where teams optimize, improve, and efficiently use **operational expenditure (OpEx)** for CD. The idea is to optimize and continuously improve the known costs posture and remove unknown costs, implementing features that provide insights through the auto-discovery of underutilized resources. Developers are under pressure to deploy applications quickly, yet in doing so, they often leave environments running in the public cloud when not in use. Cloud providers charge for these open environments, even though they're not being used. Features such as *automating, decommissioning, and optimizing runtimes to control cloud costs* are gaining momentum; companies are either starting to build capabilities or integrating off-the-shelf tools. Let us look at one other aspect:

  - **Kubernetes cost management**: Another area is Kubernetes and performing cost optimization on workloads. Software tools such as **Kubecost** make recommendations on optimizing your cloud spend. The reason cost management tools are important for cloud-based CD is that these tools help to track the cost of Kubernetes across the entire pipeline, even when the pipeline is spread across various cloud providers. It also helps to allocate costs to teams, individuals, projects, or environments. Further details are available here: `https://www.argonaut.dev/blog/kubernetes-cost-optimization-tools`.

For regulated sectors, standards and compliances are critical. Let's take a closer look:

- **Certifications and standards**: There are many organizations working actively toward the standardization of cloud technologies. Most widely referred-to standards are from ISO and NIST, and then there are industry-specific standards as well. NIST develops and distributes standards primarily for government use but these are also widely used by private industry. Examples of basic cloud compliance standards from NIST include the following:

  - NIST *SP 800-145* (2011): The NIST definition of cloud computing
  - NIST standards acceleration to jumpstart the adoption of cloud computing

- **NIST Cloud Computing Program** (**NCCP**)

- NIST *SP 500-291* (2011): NIST cloud computing standards roadmap

- NIST *SP 800-144* (2011): Guidelines on security and privacy in public cloud computing

ISO is one of the primary global standards-making organizations. Some relevant standards are shown in the following list:

- *ISO/IEC 17789:2014*: Information technology—Cloud computing—Reference architecture

- *ISO/IEC Technical Specification (TS) 23167:2020*: Information technology—Cloud computing—Common technologies and techniques

CD is still evolving, and some organizations are attempting to standardize implementation, such as the **Continuous Delivery Foundation** (**CD Foundation**) and the **Cloud Native Computing Foundation** (**CNCF**). To do this, it is important to gain an in-depth understanding of the required mindset and culture, which we will discuss in the following section.

Understanding and owning the culture for CD in the cloud

Maximizing CD outcomes in the modern cloud ecosystem comes with leveraging the productivity mechanisms offered through technological advancement. Taking the next steps in the journey and embracing the CD continuum might be more complex than it looks.

As an organization passes through different stages, it is important to continue to collaborate and share learning to continuously improve the CD posture. Here are a few points that will help you to avoid stalling your adoption journey once you are past the first few steps of setting up CD for products:

- **More automation as we go**: Evolving from removing toil to orchestrating a set of automated tasks in the workflow needs a team effort. Once the automation scales beyond a particular task, it needs substantial collaboration and coordination between various functional teams. The next steps in the automation journey for CD and the cloud together can seamlessly cross over different cloud environments yet implement the standards and processes in principle—for example, IAM, policy management, or cost management—in a unified way. Another example is abstracting the underlying environment from users by seamlessly provisioning the infrastructure.

- **Embracing challenges associated with progressive delivery**: Software engineering teams might face many challenges in the process of expediting the release cycles of features, which leads to burnout and frustration. Teams are under intense pressure to reduce their cycle times but still improve the application's quality. Moreover, an increased threat landscape and evolving regulatory and compliance practices make it extremely complex to thrive in the ecosystem without embracing the challenge and improving practices as we go along.

Progressive delivery is one such improvement proposed by practitioners for CD and continuous deployment. It is a structured approach of releasing software to a subset of users, locations, environments, or experiences and gathering feedback incrementally. Progressive delivery comes with its own set of challenges, tools, and process integration. Careful assessment is needed to onboard to progressive delivery. Progressive delivery calls for a data-driven approach. Also, a more granular feature rollout, even delivering different experiences and configurations, needs a more efficient and automated way to provision infrastructure and perform A/B testing, and—lastly—observability as one of the core features of your delivery. Some considerations for advancing CD with progressive delivery are provided here:

- Applications such as **Kubernetes and GitOps** together lay the foundation for progressive delivery and enable systems to be declarative and controlled. It can be helpful to assess whether these applications can support you in your next steps.

- We can use several tools such as **Helm**, a tool that manages Kubernetes applications, **Flux**, a GitOps agent that automates the movement of code, **Flagger**, a progressive delivery solution that allows you to manage canary releases using policies, and **Weaveworks**, which supports operating Kubernetes workloads and simplifying management of cloud-native applications.

- Platforms for progressive delivery or **feature management platforms** provide sophisticated orchestration, observability, and self-learning capabilities to reduce the blast radius of incremental releases by reducing rollback time and embedding learnings as we go along the journey. Some of these are listed here:

  - **Argo CD** unifies cloud-native applications and infrastructure updates in a similar way. It automatically syncs and deploys applications whenever a change is made in the GitHub repository (`https://argoproj.github.io/argo-cd/`). Specifically, Argo Rollouts focuses on reducing the blast radius for feature releases and provides strategies for progressive delivery.

  - **Flux** was originally developed by Weaveworks and is an incubating project under CNCF. It has similar functionality to Argo CD; however, it lacks a user-friendly web UI.

  - **LaunchDarkly** is used for feature management with automated feature toggles and provides feature life cycle management capabilities.

  - **CloudBees** is a feature management and flag management application that enables dev teams to target different users based on various attributes (`https://www.cloudbees.com/capabilities/feature-management`).

  - **Bridging the reliability gap**: With speed comes reliability challenges. Taking motivation from **site reliability engineering** (**SRE**) principles and applying it to progressive delivery would need an iterative approach as well.

- **Setting up error budgets** based on objectively measuring and putting together **service-level objective/service-level indicator** (**SLO** and **SLI**) metrics for specific products and their features can bridge the reliability gap. Budget dashboards integrated with the prioritized allocation of a budget, adjusting it through time range control, and comparing the impact produced by the spending can provide a transparent way to fund availability and performance-related debt created by aggressive feature development. Some good examples of implementing SRE can be seen at Google or GitLab.

- **CI/CD observability**: Observing CI/CD pipelines is achieved by instrumenting different CI/CD and DevOps tools to provide monitoring dashboards, alerting, and **root cause analysis** (**RCA**) on pipelines and CI/CD platforms and to instrument tools with **OpenTelemetry**, as follows:

Example Tools	Description
Honeycomb	Honeycomb, also recognized as a leader in the observability space by Gartner and similar research organizations, has several products to observe complex cloud-native application ecosystems. They have purpose-built query engines to provide a good baseline to implement full stack observability for cloud-native applications.
Buildviz (open source; built and maintained by a Thoughtworker, a Thoughtworks employee)	Buildviz provides graphs detailing runtime behavior, failures, and stability of the pipeline. The following CI/CD systems are currently supported: Concourse, GoCD, Jenkins, TeamCity.

Table 4.5 – Simple example of CI/CD observability tools

- **Business imperative of CD in the cloud**: CD's next steps are not a start-up thing. Release orchestration aligned with business objectives such as reducing the cost of every release, optimizing releases based on user segmentation to maximize ROI, increasing developer productivity through early feedback and incrementally improving features as needed, and—lastly—reducing the blast radius all contribute to successful products and increased business efficiency.

"If your business cannot do progressive delivery, your competitor will figure it out and crash you. So, figure it out." (Andrew Glover, Netflix)

For example, Netflix uses Spinnaker with feature flags for deployment automation together with canary traffic routing. For incremental and on-time health of the services, they have also implemented observability and telemetry with Alice (`https://logz.io/news-posts/alice-slackbot/`).

Shifting left – cost, risk, and security management

With changing CD, security concerns are elevated. A new threat ecosystem is in the making, and application developers are constantly looking to mitigate the risk and reduce the blast radius of security breaches. They must also transform how they implement security across their cloud-native stack. On top of this, more rapid releases and the adoption of a next-generation cloud-native approach with tools and platforms that enable applications to be built in hybrid environments have overwhelmed developers. To successfully implement cloud-native application security, enterprises should use an integrated platform approach—in Gartner's research referred to as **cloud-native application protection platforms** (**CNAPPs**). CNAPPs are an integrated set of security and compliance capabilities designed to support, secure, and protect cloud-native applications across development and production.

There are various elements of cloud-native security, right from application security to pipeline and infrastructure security. On top of this, open source components are embedded into various applications and tools used for creating the pipelines. Static and dynamic vulnerability management becomes a priority. Kubernetes security is another rising domain where there are a number of tools available for securing Kubernetes.

Many companies are also in a transition state, transforming monolithic applications to microservices-based architectures, which also adds to the complexity when it comes to security. Ahead, we highlight considerations around enabling a security-first culture with cloud-based CD:

- **Security posture management for containers**: Security for Kubernetes comes with a number of open source tools—for example, `Kube-Linter`, an open source tool that identifies misconfiguration and programming errors within Kubernetes configurations. This tool comes with built-in checks for common misconfiguration. `Kube-bench` is another tool that provides extensive static analysis. **Open Policy Agent** (**OPA**) for Kubernetes security is a policy management tool and is often known as a firewall for Kubernetes. Other tools such as `kube-hunter` and `Falco` are also popular choices.

- **Secure hybrid cloud deployments**: It is complicated to manage hybrid cloud security and might get overwhelming with evolving roadmaps and other moving parts. In general, establish a uniform set of security standards and tools that you can apply across the hybrid environment, such as following the principle of zero trust and implementing IAM, observability, scanning tools, and so on.

- **Securing an SBOM**: Securing the supply chain for software is one of the evolving areas when it comes to cloud-based CD. With the support of an SBOM, developers can perform vulnerability and license analysis, evaluate a software product's risk, and review potential risks from newly discovered vulnerabilities.

- **Vulnerability scanning**: Scan infrastructure components, container images, and serverless functions for known vulnerabilities, embedded unencrypted secrets, OSS licensing issues, malware, and configuration issues before they are deployed.

- **CI/CD pipeline security**: CI/CD security is still a new area. There is a huge threat landscape to be addressed. One of the threat matrices was discussed at *CODE BLUE 2021 OpenTalks*.

Securing applications in a cloud-based CD ecosystem comes with its own challenges. The most important part of it is that it is still evolving. The next section highlights basic yet complex security challenges for cloud-native deliveries.

Challenges for security with cloud-native CD

Once you start to analyze the security posture for cloud-native CD, it is evident that there is more work needed in the areas of standardization, interoperability, and security compliance. Here are some of the challenges:

- No single open source or commercial security tool covers all aspects of end-to-end security.

- Security tools will require portability across hybrid environments.

- Static and dynamic vulnerabilities are difficult to detect and correct with the CI of open source components into cloud-native applications. Often, these go unnoticed until a breach happens.

- Hybrid cloud environments also bring a substantial challenge to compliance management. Auditing cloud-native applications deployed across multiple cloud platforms is a complex task; often, no standardized way to do continuous audits exists.

There is a lack of standardized security processes between different public clouds. Security is a big topic to cover holistically; however, as we conclude this section, which is specifically highlighting some considerations, capabilities, and challenges with the cloud onboarding of CD, it is advised to refer to the later chapters, where we will go into more detail on technical capabilities.

Chaos engineering

As the complexity of cloud-native applications grows, **chaos engineering** adoption is also accelerating. Chaos engineering is a technique of implementing controlled experiments to identify and remove security gaps and key weaknesses of applications by deliberately injecting faults in runtime. The whole idea is to improve the application's security posture and create more robust applications. There are many popular commercial and open source off-the-shelf tools; we will mention a few here to provide our readers with a broad idea:

- **Gremlin**: This is one of the tools in the space that has evolved with time, starting from a number of services that performed controlled experiments to a unified platform with reliability testing, scoring, and automation. It serves the need of modern software teams who are constantly looking for applications and services to improve the CI/CD posture.

- **Chaos Monkey**: This tool was invented by Netflix. It was primarily used for assessing infrastructure resilience. Further, it was released under the Apache 2.0 license. This tool randomly takes down virtual machine instances in the production environment, to assess and reduce the blast radius continuously.

- **LitmusChaos**: This is another leading chaos engineering platform, primarily targeting cloud-native applications and infrastructure. It is enabled with controlled experiments that are declarative in nature. This platform also has metrics to quantify the impact of experiments, and so on.

- **Chaos Toolkit**: This is another open source tool that advocates for **chaos as code** (**CaC**). The experiments are built in YAML/JSON files that can be shared with the team. This tool also has an open API for extending it further.

There is a good blog on the comparison of leading chaos engineering tools at `https://www.royalcyber.com/blog/devops/chaos-engineering-tools/`.

This section provided readers with a baseline for *Part 2* of the book, where we will further elaborate on technical capability.

CD practice and its evolution

The maturity of a CD practice comes with time. Once the practice evolves, faster releases with fewer bugs and issues make it to production. The flexibility of on-demand, anytime deployment of new features not only increases the competitiveness of the product but also builds confidence in the development practice itself. More standardization and maturing tools and applications for CI/CD also enhance the CD services portfolio. Organizations are also looking at different ways to improve developer productivity by removing foundational capabilities with an *as a service* option. CD pipelines move down in the value chain with the maturity of the CD practice. The stages of CD practice maturity and related operating models are set out here:

- **CD with on-premises**: The first steps in building up a CD practice indeed start with understanding the benefits, which can range from just automating manual tasks to building up the basic infrastructure, applications, tools, and processes for enabling CD with on-premises infrastructure. This step is often done in-house, with a few developers trying to improve productivity in-house, not sourced from any other **service provider** (**SP**). At this stage, there is no or limited funding for the practice, and it happens in pockets in different product lines asynchronously.

- **Self-hosted CD on the public cloud**: Once CD and cloud initiatives meet at the crossroad, a convergence of these initiatives often begins, and leaders start to create a shared vision toward CD, benefiting from the cloud initiative. At this point, teams start to explore self-hosted CD, mainly infrastructure on one primary cloud environment. With time, the initiative can grow across different cloud environments, and more exploration of cloud-based CD tools is predicted in the next steps. At this stage, the environmental complexity can grow, yet the cost of ownership can be optimized, and this can result in setting up standard procedures for build versus buy, selection of tools, and so on. At this stage, the organization starts to look at delivery partners for its CD capability as the focus shifts to innovating new features rather than spending time evaluating and setting up pipelines.

- **CD as a service** (**CDaaS**): As attempts are made to standardize the CD posture, it becomes evident that more and more organizations are looking at delivery partners to build, maintain and evolve CD pipelines. A benefit of CDaaS is removing the ownership cost of creating and maintaining the pipelines so that investments can be redirected to product feature innovation. CDaaS has a business case for companies that have a mature understanding of CD and do not want to invest more than needed in figuring out *how* to create a CD pipeline; rather, they are looking to use on-demand out-of-the-box solutions to take care of it. Many product vendors offer a variety of price points that range from free tiers to *starter kits* for a full year of the service.

Major cloud providers offer integrated tools for CI and CD. **AWS CodePipeline** (`https://aws.amazon.com/codepipeline/`) provides templates to facilitate declarative deployment of the capabilities. Such out-of-the-box solutions are evolving; one such example is **Tekton**, which is an open source tool, often used to deploy applications across cloud platforms. The key advantage is being able to standardize and provide out-of-the-box cloud-native components to support applications across platforms. Let us look at some of these:

Services	Description	Benefits
CD—Create and deploy services	One-time offering where CI/CD practitioners help organizations set up CD	The CD pipeline is deployed by experts; this reduces the time of deployment and the challenge of building or selecting the right toolset for your environment.
End-to-end CDaaS	On-demand, subscription-based standardized CI/CD capability with support	Instead of adopting a separate technology for CI/CD, the tight integration of end-to-end features make this an attractive option. When it comes to pricing, the CI/CD features are blended into the subscription. GitLab and Harness are examples of services that provide this kind of subscription.
Customized CI/CD implementation	Hybrid and customized implementation	There is no one-size-fits-all tool; if you are in a unique situation where you need to customize CI/CD implementations, some system integration companies provide you with this capability.
CI/CD assessments and maturity	Identifying key strategic gaps	If you have implemented CI/CD and are looking for assessing and enhancing maturity by adding special security and compliance features or observability features, for example, this service can be an option.

Table 4.6 – CD services

- **Managed CD**: CD can be overwhelming to manage once it starts to scale. It not only creates overhead for the product development team but it is also challenging to keep costs and resource utilization under check. To solve such issues for more mature organizations, there are services that can manage your releases and the pipeline infrastructure. **Google Cloud Deploy** is one such example.

In this section, we looked at various options to integrate various types of services into the CD operating model. It is evident that there is no one-size-fits-all service(s); it depends on organizational priorities, maturity, and scale. In the next section, we will take a deep dive into the requirements for scaling CI/CD and key considerations before scaling.

Key requirements for scaling CI/CD

Once the number of microservices tends to increase, the CI/CD ecosystem can get complex very rapidly. Unfortunately, adding more developers cannot be the modern solution to this problem. The number of deployments no longer linearly scales by adding more developers to the equation. A lot has been already said about strategizing CD, but to conclude, here are a few key pointers to think about when you are scaling CI/CD for an organization:

- **Ecosystem integration**: Standardize and automate the CD pipeline. Clearly define the ownership for creating the deployment pipeline, and when performing deployment, try to automate as much as possible to reduce management overhead in a stepwise manner. Also, ensure that you can measure how current processes/tools and applications are performing.

- **Implementation of a hybrid cloud**: Whether you're running in AWS or Azure, building software in Google, or using a multi-target approach, always keep the business objective in mind. Take advantage of the key characteristics of the cloud without getting tied into using one vendor or many customized integrations.

- **Container strategy**: Remember to evaluate whether you have the flexibility to deploy the code you are building to any cluster, namespace, any or all K8s or other container environments, at any scale, in a public cloud or on-premises.

- **Observability strategy**: Always look for bottlenecks and apply the theory of constraints. Ensure to implement end-to-end observability to take a closer look at performance.

In this section, we pointed out some key concepts to consider when planning to scale CD in the cloud. Some of these aspects will be detailed in further chapters.

Summary

In this chapter, we covered various topics that are important not only for setting up your journey toward CD in the cloud but also scaling it with confidence. The sub-sections helped you understand aspects around choosing the right cloud environment and the right specific capability of CD. We also stressed the critical aspects of security. In the next chapter, we will touch base on the technical strategy.

Part 2: Implementing Continuous Delivery

This part focuses on the technical implementation of continuous delivery in the cloud. First, we'll deal with some architectural issues before walking through some technologies that can be used to support continuous delivery in the cloud. This part will then deal with security and testing strategies, before ending with some tools that can be used.

This part contains the following chapters:

- *Chapter 5, Finding Your Technical Strategy Toward Continuous Delivery in the Cloud*
- *Chapter 6, Achieving Successful Implementation with Supporting Technology*
- *Chapter 7, Aiming for Velocity and Reducing Delivery Risks*
- *Chapter 8, Security in Continuous Delivery and Testing Your Deployment*

5

Finding Your Technical Strategy Toward Continuous Delivery in the Cloud

In the previous chapters, we dealt with goals and target metrics when planning for continuous delivery. We learned that there is always a strategic goal we follow when we deliver our software, and that cost-effectiveness is an important factor for our applications. Although these are all more strategic and economic goals, they are also important when designing our technical solution. We should not only have the business impact in mind when we deliver our software but also knowledge of how we can support the business goals with our deployment strategy.

This chapter will guide you through some questions and thoughts you might have when designing your technical continuous delivery strategy:

- What am I deploying? (monoliths versus microservices)

- Where am I deploying to?

- What about cloud-native platforms?

- How many cloud providers are involved? How could I keep them converged?

- How often do I want to deploy and why?

Throughout this chapter, you will learn about all the things that might influence your technical deployment strategy, more or less independent of the technologies used (which will be covered in the next chapters). We will start with some simple things that might sound obvious to you but are often forgotten when designing a deployment strategy, discuss architectural things (such as what monoliths and microservices are), and finally dive into cloud-native and large-scale environments. With all of this under your belt, you should have a good overview of the things to consider when designing your deployment strategy.

We will cover the following topics:

- Ensuring the technical success of your deployment strategy
- Exploring monolithic and microservice architectures
- Cloud-native platforms and considerations
- Large-scale strategies

Ensuring the technical success of your deployment strategy

As in every engineering-related topic, there should be goals and a definition of success in your deployment strategy. In this section, we will give you some examples of success factors for your deployment journey. You might find out that many of them are not necessarily related to the people who are creating the CD platform:

- **Cooperating**: Deploying software should be a collaborative effort. While there are people in charge of designing the infrastructure and continuous delivery processes, there are also those who are designing and writing software. It is of vital importance that platform/infrastructure/ CI engineers and **Site Reliability Engineers** (**SREs**) understand the needs of the developers and the software to be delivered. Also, developers have to be aware of the infrastructure their software will be deployed to and which processes and platforms are involved in the process.

- **Using the right service model for your application and its components**: At the moment everyone talks about cool and fancy things such as **Kubernetes** and **serverless applications**, and it might be tempting to think all of these technologies are the response to all questions when it comes to operating applications. Always keep in mind that even the managed service offerings of such platforms might require a lot of knowledge and could cause a lot of overhead on a small scale. Opting for **Infrastructure as a Service** (**IaaS**) might be an obvious choice when it comes to shifting workloads to the cloud, although there might be managed services that allow you to focus on your applications instead of the underlying infrastructure.

- **Question current processes, be ambitious, act in small steps**: Especially when shifting workloads to the cloud, it makes sense to clean up and question classic processes and take a look at the current state of the art. Although many long-used processes have their rationales, these might be dependent on limitations that no longer exist and could be mitigated technically when delivering to the cloud. Furthermore, be aware that designing and creating a delivery system is a continuous process. Therefore, it's totally valid to have high goals for your work and determine how you want to get there. Nevertheless, tackle them in small steps and try to get there via incremental improvements to ensure business continuity, as delivering your software can be considered as a critical process.

- **Keep it simple**: All of these technologies and processes are great and can help a lot when designing a delivery system. However, it is important to keep in mind that this system has to be maintained by someone in the future, so it should be easy to understand and maintain. Therefore, keep it simple and do not over-engineer your solution. It is always better to have a simple solution that works than a complex one that does not.

While building your delivery system, you want to have something you can present to your management that indicates the value this brought to your company. We will discuss the well-known **DevOps Research and Assessment** (**DORA**) metrics in *Chapter 7*, but at this point, it's more about your goals. For instance, you have to decide whether you want to go slow or fast. If you plan to follow a more defensive approach and deploy delivery in a defined timeframe (e.g., every two weeks), you have to find more concrete answers to the question of how to bring fixes and security patches to production if it has to go fast. This might lead to additional problems such as a divergent code base (where what is checked in on the main branch is not what is deployed in production) and backporting changes. If you follow a more progressive approach, you want to push your changes as soon as possible, but in this case, you have to deal with frequent changes for your customers, ensuring the quality of your pipelines and having mechanisms to get back or forward to a more stable state if something fails. We will deal more with these issues in *Chapter 7*.

Furthermore, don't forget about your engineers when designing your delivery system. Your developers should be aware of the platform they are deploying to and the mechanisms that take place there. It is important to give them guidelines on which things they have to take care of, and—in the best case—templates to build the software in such a way that it will be highly compatible with your platform.

For completeness at this point, you should always have security in mind when designing your delivery system and should make your success measurable. We will deal with velocity and performance measure management in *Chapter 7*. Furthermore, we will deal with security in *Chapter 8*.

In this section, you got an overview of potential goals and things to consider when designing your CD strategy. In the following sections, we will deal with more technical topics and considerations for the design process.

Exploring monolithic and microservice architectures

In this section, we will discuss briefly what an application is and some traditional architectural models—monoliths and microservices—as well as their advantages and drawbacks. We will also discuss some challenges when dealing with monoliths and microservices in cloud environments.

We talk a lot about applications in this book, but as this term is a bit plastic, we should keep an eye on what an application could be for us. When we're tapping and swiping around our mobile devices, we often use the term *app*. So, in the mobile context, an **app** is a piece of software that enhances the capabilities of our device. In a more traditional sense, on our desktop/laptop, an **application** is very similar to an app and is often written in a very monolithic approach. Therefore, we call things such as our development environment an application, and things that run in the background a service or daemon.

In the context of this book and in larger systems, this behaves a bit differently. An application might consist of multiple services/components and might be highly distributed. Nowadays, we encounter two major architectures when building our applications: monoliths and microservice applications.

The **12-Factor App Methodology** (`https://12factor.net/`) describes how to build services in a way that they stay scalable and maintainable. Many of them can be applied to monoliths and microservices but are mainly targeted at microservices. For the sake of our deployment focus in this book, we will take a closer look at some of them:

- **Dependencies**: To deploy our software, we want to bundle all of our dependencies and always be able to build a reproducible environment. One of the technologies that helps with this the most is containers, as we can bundle all of the distribution packages with the software and deploy them.

- **Config**: All configurable parameters of the software should be configurable via the environment via environment variables or mounted config files. This enables us to use the built artifacts (as containers) throughout all environments and simply change the parameters needed for each environment.

- **Dev/prod parity**: It makes sense to keep development and production environments as similar as possible. This influences our technical strategy: we need to keep the time frame between the code change and deployment as short as possible to ensure that developers are still aware of what they changed when bug-fixing, and the systems to reproduce problems behave the same.

Next, let's consider a simple example to help you compare the two architectural styles:

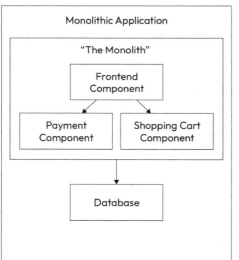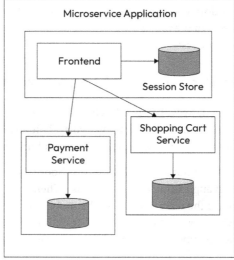

Figure 5.1 – Monolith and microservice applications

On the left side, we see a typical monolith consisting of multiple components (*Frontend*, *Payment*, and *Shopping Cart*) connected to a database. On the right side, we've got the same application in a microservice style, and the *Data Stores* connected to each service. They fulfill exactly the same purpose and might do their job in exactly the same manner; they could also have exactly the same code (with the exception of the interfaces between the components/services). So, if the two things are completely the same, what is there to worry about? We will discuss this in more detail in the following sections.

What are monoliths?

When we talk about monoliths, we're mainly thinking about the customer-facing part of an application, which includes a user interface and the backend. In many cases, monoliths consist of multiple services, such as databases, filesystems, and so on. In the past, these services were maintained by different teams and had their own life cycles (and often, people were afraid of database upgrades). Nowadays, these components should also be maintained and updated regularly and might also be deployed side by side with the monolithic application.

Monoliths are often deployed as a single deployment unit and built and tested together. These characteristics make our delivery strategy very easy at first sight, but when it comes to high availability and advanced deployment strategies, we have to consider some things:

- **All-or-nothing deployments**: When we deploy a monolith, we also always deploy all components included with it. In our example, we will have to update the whole monolith when we make a simple change in the shopping cart service. This doesn't mean that we cannot build safety nets around it. For instance, nobody is preventing us from deploying the new version of the monolith first, checking if everything is ready and then removing the old version of the monolith. Therefore, it's safe to say that the blast radius of a failed monolithic deployment can be very large, but there are also ways to avoid it.

- **Scalability/elasticity**: As we described earlier, we could create a second instance of the monolith for upgrading, but also for scaling when we need to handle load. As monoliths could get very large, this might have an impact on resource usage and, as a further consequence, the costs of our system. Let's imagine that we find out that our shopping cart component needs to handle a bit more load and we want to scale out (add an additional instance). In our monolithic setup, we will also deploy the other components that might not necessarily have to be scaled.

In general, deploying a monolith can be very easy, as you have only a limited number of dependencies and can ensure that the application is consistent in itself (as it will most likely be developed and tested together).

In the literature, it's often described as a disadvantage of monoliths that many developers are working on the same product and that the addition of new developers on such products may have a negative impact on the project's velocity. Nevertheless, many products start as monoliths, and when they are properly designed/componentized, these things might not be a big issue in the end.

Our example might also evolve, changing from a monolithic approach to a so-called microservice approach, so let's discuss microservices a bit.

What are microservices?

Generally, microservices means a collection of small services that, when combined, form a larger application. There are many approaches to separating the pieces of an application into microservices. In our case, we split them into their functions (payment, shopping cart, and frontend) to be able to re-use the code later if we need the function again elsewhere. As you can see in *Figure 5.1*, these services have their own databases/attached services, which reduces the dependencies between them. In real life, you might not necessarily deploy a full-fledged **Relational Database Management System** (**RDBMS**) for each microservice, so this could also be a separation of schemas or databases. One thing to consider in addition is that the services integrate via APIs and not over the database. If the payment service needs data from the shopping cart service, it would fetch the data via a defined API, and the shopping cart service would in turn be responsible for getting the data from its own database.

Microservices overcome many of the disadvantages of monolithic applications. For instance, each microservice (which might be a component in a monolith) can be deployed and scaled individually, which leads to more efficient resource consumption and more simplified scaling of the application. Normally, data management is decentralized in microservices, so scaling the data-intensive services can be a bit easier.

Although all of this sounds very nice and promising, there are also some challenges here:

- **Different programming language**: One of the most frequently described benefits of microservices is the freedom to use various programming languages for different services, which makes it easier to hire developers and use the language that best suits a service. This is a thing I only agree with partially, as companies might have their own preferred languages, and it makes it easier to shift a workforce between services if they are dealing with the same language.

- **Pipeline standardization**: Often, microservices are developed in different repositories using separate pipelines. Sometimes it can be challenging to keep them consistent. Therefore, templating and automatic upgrading of CI/CD systems might be a good start to avoid divergence here.

- **Bundling/Versioning**: Although microservices should be deployed independently from each other in the literature, this might not be the best approach for your application and might not fit the maturity of your deployment strategy.

As the microservices might be developed and deployed individually, they might depend on other services (or other services are dependent on them), which leads to certain overhead and considerations when testing and developing such applications. Now that we have discussed the delivery-related aspects of microservices and monoliths, we will try to sum this up and see if monoliths are as bad as they are often perceived.

Are monoliths bad, and should we always use microservices?

We discussed some aspects of monoliths and microservices here, but you might ask at this point: which should I go for? It depends. If you have an idea for an application and do not care too much about scaling at the moment, you might be good to start with a well-defined monolith, which means that you have a clear structure for your components and have an idea of how to split them into some microservices. With this approach, you will be able to transition into a microservice application when the time is right. Monoliths are not universally bad at all, and when starting on your application it might help you get up to speed very fast.

On the other side, it is a good idea to start with microservices if you know that you will have multiple teams maintaining your services and want to scale the services independently. With all of the shiny advantages microservices bring, there also come some challenges, which we discussed previously in this chapter. For instance, you will have to decide whether you want to structure your microservices in multiple pipelines/repositories or if you follow a **mono-repo approach**. Furthermore, you will have to decide how your microservices should be split up and which teams are responsible for which services. Whichever architecture you go for, always keep the organization in mind and think about the target platform to which you want to deploy, along with how you plan to do it, and make your choice on that basis – not as a dogmatic decision.

As a result, you will find out that it's not always black or white when deploying to the cloud (and this would be far too simple). You will always have to consider trade-offs that are very specific to your applications and your organization. In the next section, we will find out what cloud-native means and how we can embrace it.

Cloud-native platforms and considerations

The **Cloud Native Computing Foundation** (**CNCF**) defines a **cloud-native infrastructure** as one with "*microservices, service meshes, infrastructure as code, containers, immutable infrastructure, and declarative APIs*" as typical characteristics. We have already looked at the application basics that support embracing cloud-native infrastructure by moving from monolithic to microservices-based architecture. In this section, we will look at the other key cloud-native characteristics of infrastructure:

- **Infrastructure as Code** (**IaC**): By describing our infrastructure resources as code, they can therefore be stored in source control systems. IaC tools utilize a declarative language to describe the desired state and have the operational knowledge to know how the desired state can be reached.

- **Immutable infrastructure**: This states that infrastructure is provisioned in such a way that it cannot be modified once it's running. Therefore, changes in the infrastructure would result in a new deployment, which means that the existing infrastructure is destroyed once the new infrastructure is provisioned. This tends to be a more reliable and secure system, as the system could be operated in a read-only mode, and doesn't store changes on the running system.

- **Declarative APIs**: These are becoming popular especially in the context of IaC. A declarative API is a desired state system. You just provide a certain state you want the system to create and don't need to worry about all the steps needed to achieve that state; this approach is popularly used in tools such as Kubernetes and Terraform.

In traditional infrastructures, we had our servers pre-installed with the packages needed to run applications (e.g., web servers or application servers) and delivered the applications by hand. Over time, this evolved to include scripts doing these tasks via SSH or APIs. These processes worked very well if no configuration changes were needed in packages and no additional software was necessary. However, when configuration changes were needed, it often happened that these changes were made in the pre-production stages but were missing when applications were deployed, as the people who deployed applications to production systems were different from those who developed them in the pre-production environments.

Many of these issues were solved when configuration management tools such as **Ansible**, **Chef**, and **Puppet** got more popular and helped make infrastructure reproducible. Nevertheless, it also came down to configuration drifts between the desired and the current configuration when infrastructure changes were done by hand and not in the configuration management code.

With the shift to cloud instances (IaaS), and the related rapid provisioning and self-service capabilities, it got easier to provision infrastructure as well as applications in a reproducible way. It is also important to note that more and more tools facilitating cloud resource provisioning are gaining popularity, partially due to the reduction in the work involved and the simplification of provisioning cloud resources in hybrid environments.

Using IaC tools, it is very easy to provision and pre-configure new instances. In most cases, we could also deploy the complete infrastructure and de-provision the old one after the application has been tested successfully.

At this point, we could use the same deployment mechanisms as we had in the pre-cloud era, for example, deploying by hand or using the scripts we wrote earlier. But we could also go one step further and create/update this infrastructure automatically using our pipeline tool of choice and update the application in one step with the infrastructure. A simple example of such a pipeline could look as follows:

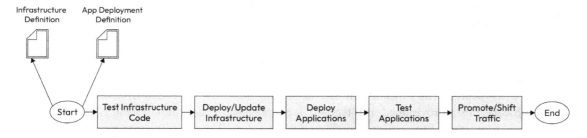

Figure 5.2 – Combined pipeline

Using such a process, everyone is working on the same goal and the outcome is deterministic. The people responsible for the infrastructure write their infrastructure code and store it in a versioned way in a repository. Application developers also maintain their code and every new version gets delivered into an artifact store and can be delivered easily to each system.

As we learned in the previous chapters, there are cloud services and delivery models that influence the way we deliver our software. The delivery models (public, private, or hybrid) affect *where* we deploy our applications and *who* is responsible for providing the respective hardware. If we are deploying our software to a public cloud, we can assume that there are almost unlimited resources available. On the other hand, in private or hybrid clouds, we are limited to the resources we have in our data centers or provisioned upfront. Therefore, spinning up an additional environment dynamically might get a bit challenging. The service model affects *what* we are deploying.

Using IaaS, we have the most flexibility in terms of how we can deploy our operations infrastructure from the operating system environment to the application environment. This changes a bit on **Platform-as-a-Service (PaaS)** offerings, where we get a pre-defined application environment and can deploy our applications as long as they are conformant to the underlying infrastructure (i.e., programming language and libraries).

Another option that has become very popular in recent years is containers, which let you deliver the runtime environment for your application and the application itself in a single artifact. The major benefits of containers are their portability, as they are the same from the developer environment to the production environment, and the density in which they can be deployed (the resource footprint is lower in comparison to virtual machines). There are multiple ways to deploy containers; they can be deployed on IaaS environments using the respective runtime, or in PaaS environments if containers are supported, but there are also native container orchestration services available in almost every cloud platform that provide the runtime/control plane for you such that you only need take handle writing your containers. There is a whole ecosystem around containerization and container orchestration (called **cloud-native**).

Kubernetes has emerged as the de facto standard around container orchestration, and a large ecosystem has developed around this. Kubernetes makes both infrastructure automation and declarative configuration easier. Kubernetes allows you to integrate various tools to automate the scheduling, deployment, and monitoring of containers. At this point, it would be way too much to describe what Kubernetes does in detail; therefore, we will stay at a very high level when it comes to Kubernetes specifics.

In a nutshell, Kubernetes is a control plane that helps you to orchestrate your workloads (a group of containers called **pods**) on a group of nodes (**cluster**). The orchestrator takes care that the pods are scheduled on the most suitable node but also inspects the state continuously and restores the desired state (number of replicas, specifications) if it no more matches. For instance, say you define that there should be five instances of one workload running. If one node running one of the instances crashes, Kubernetes ensures that this missing replica gets rescheduled on another node.

It is a widespread assumption that containers and managed Kubernetes services are cloud-agnostic, which is only partially true. As the ecosystem is very large and Kubernetes is very extensible, the platforms are not always the same and sometimes do not behave the same way. Obviously, if you are using the fundamental features of Kubernetes, they might work in most cases. When it comes to features that are provided via interfaces (networking, ingress, etc.), their availability differs between the offerings, and also the configuration might not be the same everywhere. This is very important to consider when running applications on multiple cloud providers.

There is a very large ecosystem around Kubernetes and cloud-native technologies that grows from day to day. As we are going to describe in *Chapter 6*, there are GitOps tools that help us deliver our software efficiently. At this point, we want to highlight some technologies that help us deliver our software in a cloud-native world:

- **Service meshes**: Service meshes are a way to manage communication between services. They are a layer between the application and the network. They provide a lot of features including load balancing, service discovery, circuit breaking, traffic shifting, and many more. They are also very useful for observability and security. There are multiple service meshes available, of which the most popular are **Istio**, **Linkerd**, and **Consul Connect**. In the context of continuous delivery, they are very useful when it comes to traffic shifting and canary deployments, as we will discuss in *Chapter 7*.

- **Policy agents**: When applying configurations to a Kubernetes cluster, by default, they are applied without any further checking. Policy agents, such as **Open Policy Agent** or **Kyverno**, help us to enforce policies on our clusters. They can be used to enforce policies on the configuration of our workloads, and also on the configuration of the cluster itself. For instance, we could specify that all workloads should use a specific image registry or a specific service account. They can also be used to enforce policies on the cluster itself, for instance, we could enforce that all workloads use a specific network policy.

- **Observability**: With high levels of automation, we are able to deliver our software very fast. But this also means that we need to be able to observe our systems and applications in an equally responsive fashion. There are multiple tools available to observe our systems. The most popular ones are **Prometheus**, **Grafana**, and **Jaeger**. Prometheus is a time-series database that stores metrics. Grafana is a visualization tool that can be used to visualize the metrics stored in Prometheus. Jaeger is a tracing tool that helps us to visualize the communication between services. All of these tools are very useful for observability and can be used to monitor our systems and applications.

This is only a small selection of the technologies available in the cloud-native ecosystem. The CNCF provides a landscape for such tools (`https://landscape.cncf.io/`), which is updated regularly and gets larger every day. Using these technologies, you can build standardization layers and platforms to deliver your software in a cloud-native way and avoid reinventing the wheel with every new service. We will discuss this approach to platform engineering in *Chapter 6*.

In this section, you learned that containerization is the core technology of cloud-native applications and Kubernetes is the de-facto standard for container orchestration. You also learned that there is a large ecosystem around Kubernetes and cloud-native technologies, which is growing every day. While we will take a closer look at related technologies in the following chapters, in the final section of this chapter we will examine some strategies for delivering software at a large scale.

Large-scale strategies

In a perfect world, we could assume that our software is developed perfectly and when we deliver it to our production environment, it works as intended. The applications we deliver might consist of multiple services developed by different teams. Although microservices should be de-coupled and APIs follow a clear versioning scheme, we want to test the functionality and performance of the whole system before putting it into production.

Traditionally, such tests have been realized using multiple stages. The most commonly used scheme for achieving this is using a development, quality assurance, and production environment. In the real world, there are various concepts for stages but regardless of the specifics, these concepts should be carefully planned and thought out.

In this section, we will take a look at the different stages and their purposes. We will also take a look at the different strategies for deploying applications to these stages.

Multi-stage and multi-cluster environments

Multi-stage deployments do not necessarily have to run on dedicated environments or clusters. For cloud-based development environments used for testing in continuous integration, it totally makes sense to run them on-demand on dynamically provisioned infrastructure and remove them as soon as their purpose is fulfilled. In smaller Kubernetes environments, development, staging, and production environments could also run on the same cluster in different namespaces. Against this is the fact that the failure domain is not isolated in this case (meaning that problems in development environments could impact the production environment).

Furthermore, we should ensure that the data is always isolated from the environment to avoid writing to the wrong environment. Therefore, it could be cost-efficient to run environments of the same quality (if we have multiple dev environments) on the same infrastructure. In any case, it is a good practice to split different environments (dev/staging/production) into different infrastructures and, in a perfect world, into different cloud accounts.

As we will learn in *Chapter 6*, there are techniques to give us confidence in our applications before putting them into production without additional stages/environments. These include blue-green and canary deployments, as well as feature flagging.

As discussed in the previous section, services on different cloud providers (such as Kubernetes) might not behave the same way everywhere. In any case, we might not want to rely on a single cloud provider and spread/replicate our applications across multiple ones. If this is one of the goals of your deployment strategy, you should be aware of the fact that the configuration of the services looks different on the cloud providers, therefore you might not find a generic way to configure a database on every cloud provider. The good news here is that you can create templates for your IaC tooling to achieve this.

It might get harder if the applications are using backend services that are not available on every cloud provider. Therefore, decisions regarding backend services should be taken very carefully and if you get into a situation where a service is not available on a cloud provider, you will very often be pushed into a make-or-buy decision depending on whether you want to run the service yourself (and Kubernetes simplifies this a bit) on all the clouds or if you only want to use your hand-crafted solution on the cloud providers that don't have their own offering for the feature in question. One possibility could also be to use another third-party offering. Nevertheless, when you decide to go multi-cloud, you should also ensure that this backend service is spread across multiple clouds too.

Assuming you just want to run a Kubernetes environment on multiple cloud providers, there will be managed offerings available from almost every provider, therefore their operation seems straightforward at first glance. When taking a closer look at this, it becomes much more difficult.

First, the IaC providers for managed Kubernetes environments differ a lot between the providers as the underlying infrastructure is very different. Furthermore, the interface plugins (such as the Network Interface) can also vary between the cloud providers. For instance, **Network Policies** (these are somewhat equivalent to firewall rules), are not available by default from every cloud provider as some of them use their own mechanisms, and therefore the configuration might differ between them.

Another example is the way networking connections are exposed to the outside world (different ingresses). With this, you can try to harmonize the infrastructures to the point that they behave very similarly (install similar networking plugins or ingresses). Another option is to use a Kubernetes distribution that ensures that the clusters behave the same way everywhere. Last but not least, you could consider developing your own solution.

Rolling out applications at a large scale

At the beginning of your deployment journey, you will most likely think of delivering your software to mean going through the *development*, *quality assurance*, and *production* stages. Especially when you are operating on a large-scale or multi-cloud environment, you may encounter multiple production stages. As described earlier, this can lead to issues in terms of the cloud-agnosticism of your application. On top of that, this can lead to additional considerations when delivering a new version of an application as there might be an additional layer of complexity requiring special attention. One example is shown in the following figure:

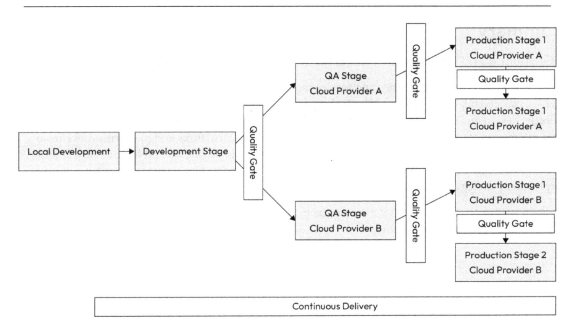

Figure 5.3 – Multi-stage environments with quality gates

Normally, developers tend to test their code locally, then after some time the code is pushed, merged, and might be pushed to a development stage. During this stage, some automatic tests should happen (ensuring that the application provides the required functionality) and after they have finished, the same artifact could be deployed to the respective cloud providers. There, more tests will occur to ensure the functionality on a more detailed level, along with load and performance tests that should be executed before deploying the application on the first production stage for this provider. There it might be useful to deploy progressively, first updating one environment, checking for errors on the customer side, and then updating the next two, and so on. Using this approach, it is possible to ensure that the application runs and is generally functional in the first instance, then that it works on the respective cloud provider as the second step, and as the third step that it provides acceptable service for the users on the respective cloud provider.

In this section, we uncovered strategies for delivering applications on a large scale, covering environments with multiple clusters and multiple stages as well as staged roll-outs.

Summary

This chapter started with a short overview of the technical success factors when delivering applications.

We learned that collaboration, the right service model, and thinking about the current processes can make us successful. Furthermore, we looked at possible metrics and found that we could measure the effectiveness of our safety nets. After that, we took a closer look at the differences between monoliths and microservices and also discussed how cloud-native infrastructures and platforms differ from traditional infrastructure. Finally, we looked at larger-scale deployments and learned how multi-stage, multi-cluster, and multi-cloud deployments can be used to give us more confidence in our deployment strategy.

In the next chapter, we'll focus a bit more on the technology and tooling side of the story and will find out how this can support our strategy.

6

Achieving Successful Implementation with Supporting Technology

In the previous chapter, we learned about many things that are relevant before we produce lines of code. In this chapter, we will learn how to implement a deployment using supported technologies. As described in the previous chapters, planning is a crucial part of a deployment strategy. As the work in the previous chapters might have led to some meetings, you might be very excited to start implementing your deployment strategy. We will cover the following topics in this chapter:

- Deployment strategy
- Continuous integration
- Templating and Kubernetes operators
- Infrastructure as Code
- GitOps
- Platforms and developer portals
- Missing parts

Drawing a big picture of your strategy

At this point, you might be tempted to open your editor and start writing code. However, now might be the best time to open your favorite drawing tool (for example, `diagrams.net`, Excalidraw, and so on), recap your earlier meetings, and draw your deployment architecture. This will help you and others get a common view of the **deployment strategy** and avoid misunderstandings. Furthermore, this can get your pathway to a successful implementation:

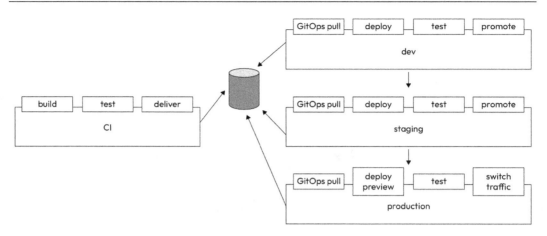

Figure 6.1 – Big picture of a deployment strategy

The preceding figure is a simple example of how to draw this. In this case, we assume that we have a CI pipeline that builds, tests, and delivers the code and finally writes the deployment configuration to a repository. This repository is read by a GitOps tool that deploys the applications to an environment. After deployment, we could execute some simple tests and promote to the next environment (for example, *staging*), where we could run some more sophisticated tests and finally promote to production. To ensure that everything works fine in production, we can deploy the new version of the application as a preview first and run some tests; once we are happy with the results, we promote the application to production. This is a very simple example, but it shows the basic idea of a deployment strategy. We introduced lots of new terms here, all of which we will deal with in this chapter.

Once you have sketched out your deployment architecture, you may have a multitude of rectangles, circles, and icons on your screen. However, the next step is to select the appropriate technologies and tools to implement this architecture on your systems. Therefore, you should start with the following questions:

- Am I already working with specific tool providers?
- Am I already part of an ecosystem?
- Do I want to use open source tools, or is it ok for me to use proprietary tools?
- Am I only working with one cloud provider?
- How open are their APIs/services? Is vendor lock-in ok for me?

After you have answered these questions, you might sense the direction you will proceed in. If you are already part of a proprietary ecosystem, it makes sense to look out for tools that are there or that integrate very well with your current infrastructure. If you are working with open source tools only, your first source for tools might be the **CNCF landscape** (https://landscape.cncf.io) or the **CDF landscape** (`https://landscape.cd.foundation`). If you want to use the tools provided by your cloud provider, the cloud provider's service catalog might be a good starting point.

If you want to work with open source tools, you have another option: proceed with reading this chapter and get kind of an overview which technologies will assist you on your continuous delivery journey and which tools will help you achieve your goals. In this chapter, we assume that you are planning your deployment strategy on a Kubernetes cluster. If you are not planning to use Kubernetes, the mechanisms described in this chapter will still be valid, but you might have to adapt the tools to your needs. Please keep in mind that every application and company works differently and the examples used in this chapter might look different in your environment.

Continuous integration

As described in *Chapter 1*, **continuous integration** (**CI**) is the process of building, testing, and packaging your code automatically in a workflow/pipeline-based way. The process is and should be triggered by every change in the code base, either by the execution of a webhook or periodic watching of the git repository. As an example, developers push changes to the respective feature branch in their code repository. The CI pipeline is triggered and builds the code, runs different levels of tests, and gives feedback to the developers. Here are some things to consider throughout the process:

- **Performance**: Developers should get feedback as fast as possible, but they want to get as much confidence in their code as possible. Therefore, it is instrumental to decide which tests should run at which time in the pipeline and how long it takes to execute the tests. Furthermore, it is always important to define the scope of each build. For example, there might be multiple components in one repository, and if one component is changed, it might not be resource-efficient to build all the components of the repository. In any way, realistic execution times should be defined for the pipeline and the execution time should be monitored.

- **Dependability**: When waiting for results, developers should be able to rely on the results of a pipeline run since wrong results in a pipeline might lead to time wasted finding bugs in the built software. Therefore, tests that fail randomly should be fixed or replaced by more reliable tests in a very short time.

- **Common effort**: One of the coolest things I have witnessed in my career is a company that fully automated its CI pipeline, allowing every developer and aspect of the process to be built seamlessly within a single pipeline. Given this, every developer saw the status of the pipeline, but it also was a common effort to make the pipeline green. This was a very good way to get developers to work together and keep the pipeline clean. Therefore, a CI pipeline is not a process solely maintained by dedicated DevOps/cloud engineers – it should be owned by the whole team.

- **Gamification**: Another cool thing I heard about was a screen in each developer's room that showed the rankings of the most stable services and the ones with the best code coverage. In this case, this leads to kind of competition between the service teams and should help raise awareness for getting things more stable

- **Security**: During the build of the software, many tools are executed and many artifacts are produced. The impact of compromised artifacts is very high, so it is very important to operate this environment securely.

There is an endless list of things and lots of interesting books about CI and as per its definition, the CI process would end after building, testing, and merging the current state of the code base. Over time, the CI process has evolved to also include the delivery and, in some instances, deploying software, which is often called CI/CD. In the examples in this book, the CI process will be building, testing, packaging, and storing the artifacts. The deployment will be done in other tools.

Now that we've learned a bit about CI, we can start to take a more detailed look at such a pipeline. A typical CI pipeline looks as follows:

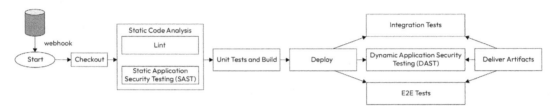

Figure 6.2 – Example pipeline

At first, the pipeline is triggered by a webhook of the Git repository. Afterward, code-level checks can be executed, such as syntactical checks (**linting**) and **static application security tests** (**SASTs**). After that, unit tests are executed and the code is built. After the build, we can try to deploy the software to a test environment and run integration and end-to-end tests. Furthermore, dynamic application security tests can be executed. After all these tests, we can assume that the software is ready, and we can store artifacts in a repository. Finally, the developers can be notified about the results of the pipeline.

The following are some typical artifacts that can be produced by a CI pipeline:

- **Binaries/packages**: Binaries or packages of software that can be deployed on a machine
- **Container images**: Container images might run on a container runtime or a Kubernetes cluster.
- **Deployment configuration**: Configuration files, which can be used to deploy the software on a Kubernetes cluster or other platforms

Normally, a **CI pipeline** consists of some entities that are responsible for the execution of the pipeline. These entities are as follows:

- **Pipeline/workflow**: The pipeline is the definition of the workflow's entire operation in a CI/CD tool.

- **Pipeline/workflow run:** A pipeline run is one instance of the execution of a pipeline – for example, the execution of a pipeline triggered by a push to a feature branch.

- **Stage/lob:** A stage is a group of tasks that are executed sequentially.

- **Task:** A task is the smallest unit of work in a pipeline. A task can be a script, a command, or a tool that is executed in a container – for example, a test execution.

The terminology of the tools might differ, but the concepts are often the same. If you are not familiar with the terminology of a specific tool, you can always look it up in the documentation of the tool. There are many tools out there that can be used to implement a CI workflow, such as Tekton, Jenkins, and Argo Workflows.

Even though CI and **continuous delivery/continuous deployment** are two different disciplines, CI is a very fundamental part of your CD strategy as this might be the entry point into your CD world. As we might have not only one service/repository to deploy and some configuration tasks might be the same throughout all of our services, we might want to create templates for our deployments. This is where templating comes into play. We will take a look at this in the next section.

Templating and Kubernetes operators

On most modern platforms, we might have some kind of declarative configuration to describe the desired state of our applications. In the easiest case, this could be a YAML or JSON file. In such a configuration, we could find things such as the name of a platform object, the container that should be started, the number of replicas, and so on. In Kubernetes, a typical deployment configuration looks like this:

```
apiVersion: apps/v1
kind: Deployment
metadata:
  name: my-cloudy-app
spec:
  replicas: 3
  selector:
    matchLabels:
      app: my-cloudy-app
  template:
    metadata:
      labels:
        app: my-cloudy-app
    spec:
      containers:
      - name: my-cloudy-app
        image: my-cloudy-app:1.0.0
```

You can find more information about deployments and other Kubernetes primitives in the official documentation (`https://kubernetes.io/docs`). In a nutshell, the preceding configuration specifies that we want to create a deployment object called `my-app` and that we want to have three replicas of the `my-cloudy-app:1.0.0` container. A controller inside the Kubernetes cluster is aware of such a configuration and will ensure that exactly three instances (replicas) of this container are running. If one of the instances is not running anymore, the controller will create a new instance to ensure that the desired state is met.

Configuring our objects this way is very convenient as we always get an overview of the current configuration. However, such configurations can become very complex, extensive, and hard to read. Imagine that you have such a configuration for a deployment, a service, a config map, a secret, and so on. Also imagine, that you have multiple environments and environment variables, or other things change between the environments. In that case, it will get very hard to update every environment, keep the environments in sync, and keep track of the configuration. This is where **templating** comes into play.

At the time of writing this book, there are two familiar templating tools for Kubernetes out there: Helm and Kustomize. Let's take a short look at them.

Kustomize

Kustomize is part of the Kubernetes project, works with overlays, and doesn't need its own templating language. Therefore, it's a good choice if you like to work with Kubernetes manifests and don't want to learn a new templating language. The following figure shows a typical Kustomize configuration:

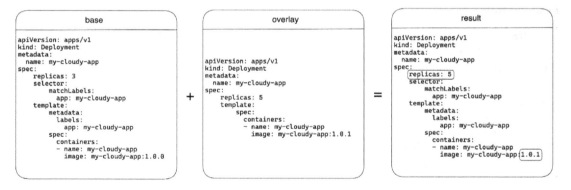

Figure 6.3 – Kustomize

The preceding figure shows a simple example of the operation of Kustomize. The base configuration can be a manifest, as we defined earlier. The overlay configuration specifies which parts of the configuration should be changed. In our example, we have changed the number of replicas and the image tag. The missing link to merge these two configurations is a `kustomization.yaml` file, which includes instructions on how to merge the configurations and potentially apply other transformations. For instance, you could use a `kustomization.yaml` file to add a label to all resources or add a prefix to all resources. After applying the overlay configuration, we get the desired result.

Helm

Helm is a Kubernetes package manager and a graduated CNCF project. You could compare Helm to a package manager such as apt or yum, which helps you install and upgrade software on your system. It has a templating language that is very similar to Go templates. As Helm allows you to configure almost everything in your deployment, it's a good choice if you want to have a lot of control over your deployments. This autonomy comes with high responsibility as changes to your deployment might have a big impact on your application. The following figure shows a typical Helm configuration:

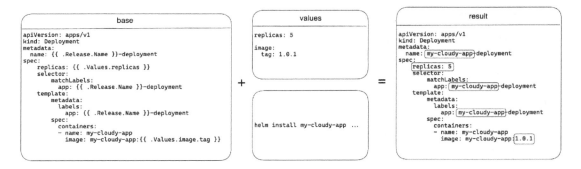

Figure 6.4 – Helm

Although this example is very simple, and there are many additional best practices to consider (`https://helm.sh/docs/chart_best_practices/`), it should give you a rough overview of how Helm works. Packages in Helm are called **charts** and are stored in a repository, which can be a local directory or a remote repository. The chart contains files that typically describe Kubernetes resources. The right-hand side of the figure shows a typical Helm template. The template is very similar to the manifest we used at the beginning; in this case, we only replaced some things we want to configure with variables. There are two sources for our variables here: **release information** and **values**. Values describe the configuration of our chart, while release information contains information about the release, such as its name. After the template is rendered, we get the desired result. When we want to install the chart, we can specify the values we want to use, either in a values file or at the command line. Finally, we can store the chart in a repository, which can help us distribute it. If you are searching for public Helm charts, you can find many of them at `https://artifacthub.io/`.

Operators

Even though Helm also has functionalities to trigger jobs before and after installation/upgrades, your application might have a more complex life cycle than just installing and upgrading. This is the point where **Kubernetes operators** can help you. Operators are Kubernetes controllers that reduce the imperative management of your application (as noted here: `https://www.cncf.io/wp-content/uploads/2021/07/CNCF_Operator_WhitePaper.pdf`). Operators can be written in various languages, with the most common ones being in `Go`; the most popular frameworks are the **Operator Framework** and **Kubebuilder**.

An operator typically introduces new **Kubernetes custom resources**, which means that you can define your own objects in Kubernetes. The controller of the operator is responsible for the life cycle of these objects. In our case, we could define a new **custom resource definition** (**CRD**) called `MyCloudyApp`, which might look as follows:

```
apiVersion: cdincloud.com/v1alpha1
kind: MyCloudyApp
metadata:
  name: my-first-app
spec:
  size: 3
  image: my-cloudy-app:1.0.0
  expose: true
```

The controller for this object type would be responsible for the underlying (secondary) resources, such as **deployments** and **services**. In this case, it would ensure that a deployment with our container is running and that a service is created to expose the application. The controller would also ensure that the state defined in the object is met, so if you change to `expose=false`, the controller would delete the service. The following figure shows how an operator (and the most declarative systems) works:

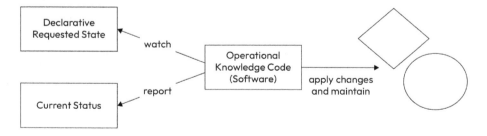

Figure 6.5 – Operator pattern

There's a piece of **operational knowledge code** that compares the declarative requested state (for example, our manifest) with the current state (for example, the status section of the object in the control plane), knows how to deal with the delta of these states, and report this back to the current state. This pattern will be used in many places in this book and is a very powerful concept.

Now, at this point, we know about three templating mechanisms, and you might be wondering which one to use. In any case, it depends on your preferences and your requirements. If you simply want to have some templating around your Kubernetes manifests, it might be the easiest option to use Kustomize. If you need some more sophisticated templating, might have dependencies between applications/ services, and want to have a comfortable way to configure your application, Helm might be a good choice. If your application has a more complex life cycle or you want to bundle your deployment with the operational knowledge to run it, operators might be the way to go for you. In the meanwhile, a combination might be the best choice. For instance, you can also use Helm to install your operator, and then use the operator to deploy your application. On the other hand, some operators use Helm to install their applications. In the end, it's up to you to decide which tool to use, and which tool to combine with other tools.

In this section, we learned about some templating mechanisms that can help us deploy our applications at a large scale and make them highly configurable. In the next section, we will take a step back and will take a look at how we can manage our underlying infrastructure as code.

Infrastructure as Code

Now that we know how to package our application, we want to get this running on cloud infrastructure. In a time not so long ago, we had to manually create the infrastructure for our application, mount servers, and configure this. As one of the major characteristics of cloud-based systems is the availability of APIs for almost everything, this is no longer necessary. We can assume that the cloud provider provides enough resources that we can use for our application. In this section, we will take a look at how we can use these APIs to create our **Infrastructure as Code (IaC)**.

First, we need to find out what IaC is and what it does. Normally, IaC is the *art* of describing infrastructure in a declarative configuration. In general, most tools that use declarative configuration work the same way as Kubernetes operators (we talked about them in the previous section). Therefore, we define our configuration in a *declarative* format, and the tooling compares this configuration with the currently stored state of the infrastructure. If there are differences, the tooling knows how to deal with these differences and how to get to the desired state. Finally, after our configuration has been applied, the new state is written back to the status information of the tool, and the credentials to log on to the infrastructure are provided in one way or the other. There are tools out there that have a *semi-imperative* configuration and do not store a state. In this case, it is more important to take a look at idempotency and ensure that every run of the tooling leads to the same result.

There are some things to consider when writing IaC:

- **Think and work like a developer**: Even if you are dealing with infrastructure, you are working with code. Therefore, you should write tests, use version control, and do the same things you would do when writing an application. Many IDEs also have plugins that support you in writing your infrastructure code, such as the Terraform plugins for IntelliJ or VSCode.

- **Use the same mechanisms as for your code**: In this case, you should store your code in a version control system and use pull requests and pipelines. This gives you more confidence in your code and makes it easier to collaborate with others.

- **Keep your code simple and don't repeat yourself** (**DRY**): IaC can save you hours of work, but it can also cost you lots of time, money, and reputation. For instance, you can modularize your code (or use templating) to make it reusable when creating multiple instances of the same infrastructure. Furthermore, IaC tools might look very simple at first and it might be tempting to use all the available features. Sometimes, and especially in troubleshooting situations, it's easy to get lost in the complexity of your code. Therefore, keep it simple and only use the features you really need.

- **Keep an eye on your tests and plans**: You are dealing with infrastructure and if the infrastructure is not available, lots of applications might not be available. Therefore, take a close look at the tests and plans of your IaC tooling. They might help you get an idea what of will happen. If some resources are changed or deleted, you should be aware of this and know the impact of these changes.

There are many tools out there that can help you write IaC. The following table shows some open source ones:

Features	Terraform	Pulumi	Crossplane
Configuration language	HCL, JSON	JavaScript, TypeScript, Python, Go, .NET, YAML	YAML
Supported resources	1,000s across AWS, GCP, Azure, and more	~120 provider packages	~70 from Kubernetes providers
Deploy to multiple locations	Yes	Yes	Yes
Primary function	Provisioning	Cloud engineering	Control plane
Paradigm	Declarative API wrapper; programmatic with CDK	Programmatic and declarative	Declarative with continuous reconciliation
Ecosystem maturity	High	Medium	Relatively new
Pricing	Free, with a paid managed offering	Free, with a paid managed offering	Free, with a paid managed offering

Table 6.1 – Comparison of IaC tools

Terraform is the most popular one and is maintained by HashiCorp. It uses a simple, declarative language called **HashiCorp Configuration Language** (**HCL**). The tool takes care of the dependencies of the used modules and supports state management as a file or external (as in cloud storage buckets or Consul). To avoid code duplication, Terraform supports the use of modules, which can be used to create reusable components.

Crossplane is a cloud-native solution for creating platforms and provisioning infrastructure. It utilizes Kubernetes Objects to store the state of the infrastructure and uses the Kubernetes API to manage the infrastructure. It is also possible to use Crossplane to manage Kubernetes Clusters. Because of its nature, it is also possible to use Crossplane in combination with GitOps controllers, as we'll see in the next section.

Pulumi is a multi-language tool that supports Typescript, Python, Go, C#, and Java as programming languages. It is also possible to use Pulumi to manage Kubernetes Clusters. Furthermore, it is possible to use Pulumi to manage infrastructure in multiple clouds at the same time.

In this section, we covered the basics of IaC for completeness. The more interesting part is the dependencies between application and infrastructure deployments. For instance, your application might depend on a database at your cloud provider or you manage storage classes in Kubernetes in your IaC tooling and have dependencies between your applications and the storage classes. Although it is possible to use the same tooling for both, many efforts are being made to establish a discipline called platform engineering. Before we take a look at this, we will look at delivering applications in a more automated way – called GitOps.

GitOps

In the earlier sections, we talked about how applications can be packaged and how we could create the underlying IaC. At this point, we might rely on the manual execution of the deployments (hopefully not) or on some kind of CI/CD pipeline. Although this is a good start, there is a smarter way to do this. This section deals with **GitOps**, its principles, and how we can use it to deploy our applications.

When we think of the things we learned until now, a typical CI/CD pipeline (simplified) might look like this:

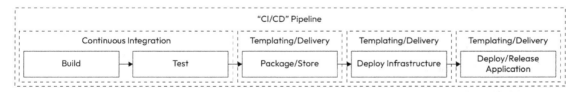

Figure 6.6 – Overview of the current deployment process

In the beginning, we build our code, test it, and store the artifacts in some repository; furthermore, we create a deployment configuration and store this in some kind of repository. After that, we create the infrastructure in a codified way and are ready to deploy our application using the respective command. As we saw in the preceding figure, we could put all of this into a pipeline, and everything is fine. Right?

Today, we are living in a very dynamic world; applications are changing very fast and so is the infrastructure. Think about the following case: you want to deploy your application in a new region. Someone makes manual changes to the target environment (for example, deletes a resource, although we hope that this is avoided). You simply want to change the configuration of your application, as well as many more such cases.

In our pipeline-based approach, we would have to change the pipeline for an additional region and the pipeline could become very complex. The same goes for configuration changes. These things are not big deals but might be annoying and time-consuming. But what if someone deletes a resource? How would we detect this and how long would it take? Remember that the pipeline needs a trigger to run. In this highly dynamic world we are living in, we need a reliable way to ensure that our infrastructure and applications are always in the desired state. This is where GitOps comes into play.

GitOps is a way to manage infrastructure and applications in a declarative way. The main idea is to have the configuration in a repository, which is the source of truth. This configuration is then applied to the infrastructure and applications. In a nutshell, this is the same pattern that we saw for operators and IaC. The **CNCF GitOps Working Group** created a few principles that define a GitOps system (according to `https://opengitops.dev/`):

- **Declarative configuration**: The desired state of the managed resources is expressed declaratively. This means that a declarative configuration (as we know it from, for example, *Kubernetes* and *Terraform*) is always our goal when using GitOps.

- **Versioned and immutable**: The configuration is stored in a system that allows versioning, which does not necessarily have to be Git. For instance, *Helm charts* could also be stored in a versioned way in a Helm or OCI registry. In any case, we want to have a way to version our configuration, which could be via tags or commit hashes in Git, but also version tags in a Helm chart or OCI registry. We should be able to rely on the fact that the configuration is immutable, so we might want to set up proper tag protection rules or immutable tags in registries to ensure that a certain version of a configuration is not changed.

- **Pulled automatically**: The respective GitOps controller should automatically pull the desired state from the source. This means that we do not have to trigger the deployment manually, as would be the case in pipelines, since the controller takes care of this.

- **Continuously reconciled**: As discussed previously, the desired state should be reconciled continuously. Therefore, it observes the current state and is always trying to apply the desired state. This is the point in GitOps where we can detect changes in the target system (*drift detection*) and automatically recover the system (*self-healing*).

As a result, the deployment process in GitOps is triggered by a push to a Git repository. This also means that we should take good care of this repository and its configuration, and it can be seen as a best practice to enable pull requests, automatically run tests, and have a proper review process. One of the drawbacks of this approach is that our *typical* CI/CD pipeline will end after the configuration is pushed to the repository (if this is part of it), and finding out if the deployment was successful is not that easy. Nevertheless, we can take this for granted and consider the CI and deployment process as separate processes that are not tightly coupled. If this is not the case, it can be a solution to make the resources identifiable by their configuration (for example, by using labels) and then use the Kubernetes API to check if the resources are in the desired state.

At the time of publishing this book, there are two major graduated GitOps projects in the CNCF landscape: Flux and ArgoCD. It is completely dependent on your preferences regarding which tool you would like to use as both tools are powerful and have a lot of features. Although they work in a very similar way, there are some noticeable differences:

- **ArgoCD**: Argo is a set of tools for Kubernetes-native workflows, from CI to CD. ArgoCD is a popular GitOps controller, which is designed for lots of use cases. ArgoCD introduces applications resources in Kubernetes, which describe the path to a repository and the configuration of the contained resources – for example, for Helm charts. The tool has a clear user interface but can also be configured via a CLI. It also has a concept called ApplicationSets, which helps generate lots of applications from various generators.

- **Flux**: Flux is also a GitOps operator with a very small footprint. The resource model is a bit different than the one in Argo as Flux uses a GitRepository resource for the repository and a Kustomization or Helm resource for the configuration. As part of its toolbox, Flux has a very powerful image automation controller, which allows you to automatically update the configuration in a repository based on images.

In a GitOps-managed system, every deployment is triggered by a change in the repository. This is the same for staging processes in multi-stage environments. Once you've installed an application in one stage and want to promote it to the next stage, you would have to change the configuration for the next stage in the repository and apply it. To some extent, this could be automated, but as this is very opinionated, we will not go into too much detail here. In any case, you might want to watch the outcome of the deployment in a certain stage, and when it is successful, you can trigger a job to promote the application to the next stage. If you are watching the state in your pipeline-based system, you could also trigger the promotion from there.

Now that we know about IaC, templating, and GitOps, we can try to combine this infrastructure and application management in a unified way. In the following section and *Chapter 7*, we will deal with a newer discipline called **platform engineering**, which helps us deal with these topics.

Platforms and developer portals

Now that we have a thorough understanding of many technologies, you might be wondering how you could use them in the real world. You may also have some other questions, such as the following:

- How can I provision lots of environments?
- How can I manage this whole infrastructure?
- Can I combine infrastructure and application development?
- What about multicloud?
- What about the developer experience?

While many of the aforementioned technologies emerged in the last years and many other people had the same questions as you, a new discipline called **platform engineering** emerged. To make this a bit more clear, let's take a look at a short example.

Let's assume the service we packaged in *Chapter 5* gets a very popular service, and we want to offer it to our customers. At some point, this application will get extended to use a database and for resiliency reasons, we want to run it across multiple clouds. Therefore, the developers of this simple service get confronted with some tough questions:

- Do I have to take care of the database for each cloud provider?
- What about all of the infrastructure services, such as load balancers, DNSs, and so on?
- How can I make sure that my application is always running?

One option would be that the developers take care of all of these things on their own, write a bit of infrastructure code for each cloud provider, and manage all of the infrastructure on their own. Another way to deal with this – especially when multiple services are involved – is to create a platform that takes care of all of these things. This platform could be used by the developers to deploy their applications while the platform takes care of the rest. This is a very simplified example, but it shows the basic idea of a platform. The following figure shows a rough overview of what a platform could look like:

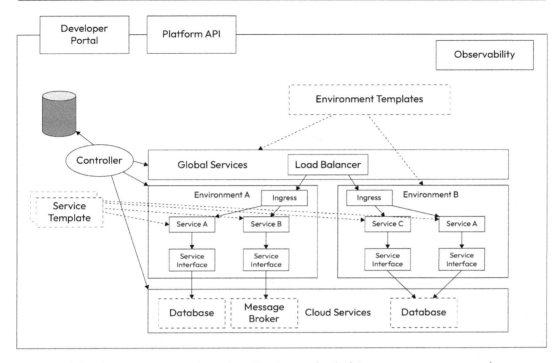

Figure 6.7 – Big picture of a platform

Before we dive into this figure, let's take a short look at the definition of a platform:

> *"A platform is a layer that provides common supporting capabilities and services for many applications and use cases. Such a platform provides consistent user experiences for getting, using, and managing its capabilities and services, including web portals and pages, scenario-specific code templates, automatable APIs, and command-line tools." (CNCF Platforms WG, 2022)*

To get back to our figure, we see some familiar things in there. First, we can see cloud services such as databases and message brokers, which can be directly consumed by the cloud providers, but also operated as services inside the platform. In any way, corresponding interfaces should exist inside the environments to make it possible for the developers to use these services. Furthermore, we can see a repository and a controller. These things will help us provision our infrastructure and services, for example, as we learned in the GitOps section in *Chapter 5*. Additionally, it makes sense to use templates for services and environments, which helps us create a consistent way of deploying these things. Using these templates, it becomes very easy for developers to create new services and environments as they only need to fill out some variables.

With your newly gained knowledge about GitOps, you might be wondering what provisioning such infrastructures could look like. In a nutshell, it could look like this:

1. A platform engineer creates a template for an environment that contains the following things:

    *   A description of the environment (instance type, region, nodes, and so on)
    *   The core services needed to get to application deployment (for example, a GitOps controller and the link to the corresponding repository)
    *   The interfaces provided by the environment (for example, custom resource definitions)

2. In the next step, this template can be used by the platform engineer to create a new environment, which will be provisioned in the cloud.

3. Once the environment has been provisioned, the GitOps controller takes care of the rest and deploys the core services.

4. Finally, the application on the newly deployed environment can be accessed by the end users. Developers get insights about the application's behavior via the developer portal.

For applications, the process is very similar. The application developer creates a new service based on templates, the deployment configuration gets pushed to the repository, and the controller takes care of the rest. In a perfect world, the developer only uses interfaces provided by the platform and does not need to know about the underlying infrastructure. This is a very simplified example, but it shows the basic idea of a platform.

Developer portals provide a consistent way of interacting with the platform. They can be used to get insights about the application behavior and gain access to the application logs, underlying infrastructure, the application metrics, and many more. The most recent emerging project in this area is **Backstage**, which is a developer portal for managing and operating software. Given its large ecosystem and the amount of available plugins, it's very powerful.

In the previous sections, we got a very good understanding of the underlying technologies that we might need for continuous delivery in the cloud. As it is often seen in the industry, we'll round this up and try to find out what's missing in all of these technologies and when it makes sense to write your own tools.

Missing parts

In this final section of this chapter, we will shed some light on the missing parts of your delivery strategy and how to deal with them. As you might have noticed, there are lots of technologies and tools out there, all of which have been built based on the ideas and maybe opinions of the people who created them. As the one who designs and/or implements the delivery strategy of your project, you might have a different opinion that doesn't have to correspond with the tool developer's opinion. Therefore, it's natural that the solutions provided might not fit your needs. Here are some examples of such cases:

- You want to make the developer experience of your platform as easy as possible and you know about the different templating mechanisms, but you are not happy with them

- You don't like the opinions people have about promoting applications between stages and want to have a different approach

- You are not happy with the way the tools in your stack are dealing with service dependencies

In these cases, you might be tempted to draw your own conclusions (it's always a good idea to draw something), open up your IDE, and start writing code. But before you start writing code, there are some things you should take into account:

- **Maintainership**: If you start writing this code, you or your team will likely be responsible for maintaining it. If this code contains bugs, you might be the one who has to fix it and if new features are needed, you might be minimally involved.

- **Dependencies**: In most cases, you write `glue-code` that connects different tools. These tools change and if they do, you might also have to change your code.

- **Existing projects**: If projects exist that are solving almost the same problem but in a slightly different way than you want, it might be a good idea to contribute to these projects instead of writing your own code. This way, you can benefit from the existing code and the community around it.

- **Talk to others**: Maybe you are not the only one who has this problem. If you talk to others, you might find out that there are other solutions for your problem or other people who have the same problem as you. This way, you can benefit from the knowledge of others and maybe even find a solution together.

Many popular tools started with a problem that the developers had and wanted to solve. Maybe your project might be the next great one.

Summary

In this chapter, we talked about the different technologies you can find in a typical delivery strategy. We started with a CI pipeline, which builds the necessary code and delivers it to a repository. After that, we talked a bit about templating and how to use it to parameterize our applications. Once we knew how to package our applications, we learned that we can define our IaC and how this can help us when deploying large-scale systems. Then, we moved on to GitOps and found out that it helps us keep our applications in their desired state, which we define in a declarative way. Afterward, we took a closer look at platforms and how to enhance the developer experience with them. Finally, we found out that not every problem can be solved and that we might want to contribute to existing projects or even write our own tools.

In the next chapter, you will learn how to be more confident in your deployment process and get faster in a controlled way.

7

Aiming for Velocity and Reducing Delivery Risks

In *Chapter 6*, we learned about the technologies and tools we can use to implement our **continuous delivery (CD)** strategy. There are lots of technologies that will help you deliver your software in a more efficient and reproducible way. So far, we have only talked about the technical aspects, but not much about how we could get faster and at the same time reduce delivery risks.

At the beginning of this chapter, we will discuss why velocity is important, but also the associated risks when getting faster. We will also discuss how this can be measured and what **DevOps Research and Assessment (DORA)** metrics are. After that, we will take a close look at things that can hold us back from our goal and how we might overcome such issues. Then, our path leads us toward deployment strategies—such as blue/green, canary, and rolling updates—and how they can help us achieve our goals. Then, we will learn how feature flags and chaos engineering relate to CD and will get a glimpse at some tools that can help us implement these strategies. Finally, we will learn how to put all of this together and how to implement a deployment strategy in Kubernetes.

We will cover the following topics in this chapter:

- The role of velocity in CD and associated risks
- Why is velocity important?
- Observability
- Deployment strategies
- Deploying and releasing

Let's first look at what velocity is and why it is important.

The role of velocity in CD and associated risks

Deployments are events on a system that can have a huge impact on it. In traditional infrastructures, this thesis led to the assumption that it's safer to deploy in larger batches and less often. Such deployments were called **Big Bang deployments** and were often done in the middle of the night. This was done to reduce the risk of failure and to have a higher chance of success. Let's inspect this a bit:

- **Larger batches**: If the changeset of software gets larger, the risk of failure and misbehavior increases, and our options when a change has a negative impact on customers get fewer. When a problem in a larger batch occurs, we might have a long changelog to investigate, and it might get very hard to find the root cause of the problem. Furthermore, it might be hard for the developers to recapitulate how they implemented a feature, as they may have implemented it a long time ago. Last, but not least, it might become difficult to create infrastructure to reproduce the problem, as the development infrastructure may have changed a lot since the most recent deployment (Dev/Prod parity).

- **Deploying in the middle of the night**: When deploying in the middle of the night, we can hide problems in our software and infrastructure from the customer. When deploying while nobody is using the system, it might not be a problem if the system is not working properly. We might not know how our software behaves under a higher load and might expect problems the next day. This leads to unnecessary heroics in the morning and dissatisfied customers.

- **Deploying less often**: Gene Kim, Jez Humble, Patrick Debois, and John Willis describe in *The DevOps Handbook: How to Create World-Class Agility, Reliability, and Security in Technology Organizations* that the more often processes are repeated, the more likely it is that they will be improved. When deploying infrequent and large batches, we will not have many opportunities to improve such problems and will always have to concentrate on firefighting when delivering software. As a result, such processes will result in a higher risk and more anxiety at the software delivery stage.

In the last few years, we have seen a shift in the way software is delivered and deployed. There are many books—such as *Continuous Integration: Improving Software Quality and Reducing Risk* (Paul Duvall, Steve Matyas, and Andrew Glover), *Continuous Delivery: Reliable Software Releases through Build, Test, and Deployment Automation* (Jez Humble and David Farley), and the aforementioned *The DevOps Handbook: How to Create World-Class Agility, Reliability, and Security in Technology Organizations* (Gene Kim, Jez Humble, Patrick Debois, and John Willis)—that describe how to deliver software in a more efficient and reliable way. Therefore, we will not go into detail on the benefits of these books but rather focus on the benefits of velocity and how we can achieve it in cloud-based systems.

Why is velocity important?

By delivering our software more often, we can provide lots of benefits to our customers and our business. In this section, we will discuss some of these benefits. When building features for our customers, we always have a vision in mind of how the feature will be used and how it will benefit the customer.

Along the way, there are many technicians, but they might not be the ones who are using the feature. As a result, the following things might happen:

- **The feature is technically perfect, but not intuitive**: Historically, some things were built in a technically perfect way but were not intuitive for the user and therefore failed.

- **This was not what the customer wanted**: Customers might not know exactly what they want, and when asking them, they might not be able to describe it. Therefore, a feature could look like it was built for the customer, but in reality, it was not what the customer wanted.

- **The feature is used in a way that was not intended**: People are creative. Maybe the product team designed a happy path to use this feature, but a customer might use it in a completely different way. This might lead to problems, as the feature was not designed for this use case.

These are only some examples, and there could be lots more of them. Now, let's get back to what we discussed at the beginning of this section and assume that we are delivering the software to the customers very infrequently. In such a case, we might not be able to get feedback from the customer on how they are using the feature and might have wasted months of development time. When delivering a long-awaited feature in such a way, we could also lose customers as they might have switched to a competitor in the meantime.

Feedback from customers can make a difference between your product and others. Therefore, velocity in software delivery can be a huge competitive advantage. With some deployment strategies we will discuss in this chapter, you will also be able to let people choose which feature sets they want to use. This will allow the customer to choose the features they want and not have to wait for the next release to get the features they want. This will also allow you to get feedback from your customers on which features they want to use and which they don't want to use.

How velocity reduces delivery risks

Apart from these business-related benefits, there are also technical ones when raising the velocity of your delivery.

To get back to the example at the beginning of the section, we are able to mitigate some risks of deploying when making this a routine process. When deploying smaller batches more often, the number of changes will be manageable, and this can help us find problems faster. Furthermore, as our development environments will still be very similar to production environments, we will be able to reproduce problems faster.

Some years ago, I heard a talk at a conference where a speaker said that his boss asked one of his juniors if he had "*destroyed something on the system today*" and got the answer "*No.*" The boss then asked, "*Why not?*" I think this is a good example of how we should think about systems and our environments today. We should not try to build our processes around possible failures but should design our systems for failure. Therefore, we should always assume that something will break, but we should always be prepared for it. This is why we should design our delivery process to take place at a peak time of the day (as a feature could be finished at any time of the day). Therefore, we will build safety nets around it; the deployment will fail once, maybe twice, or a third time. But in the end, you will be able to deploy your software at any time of the day and will be able to react to failures in a fast and reliable way.

Over time, our process will improve and we will be able to deploy at all times of the day, feature by feature. These improvements should be measured and should be used to improve the process even further. This will allow us to deliver software in a more reliable and faster way.

Measuring CD performance

When adapting your strategy to more frequent delivery, you might also want to measure the impact of your changes. Similarly to the applications themselves, you might want to adapt your delivery strategy in atomic steps, measure the impact, and then adapt your strategy again. With this approach, you will be able to find the best strategy for your business and your customers.

There are many things we can measure when it comes to CD performance; when adding **continuous integration (CI)** to the equation, these increase further. In the following diagram, you can see some of the locations where we can measure the performance of our delivery strategy:

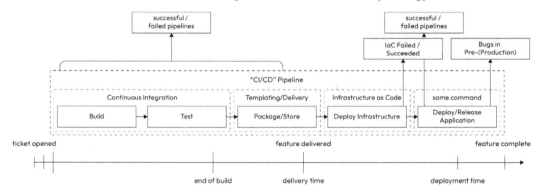

Figure 7.1 – Measurement points

In the middle of the diagram, you see a typical CI/CD pipeline, as we already discussed in the previous chapters. In the lower part, there is a **timeline**, which shows the different events we could measure. In the upper part, we see some metrics we could measure. Now, let's start with the timeline.

Normally, each feature is defined with an issue in your issue tracking system. After some time, the assigned team starts to develop the feature and commits code to the repository. One of the first things we could measure is the time between the opening of a ticket and the first commit. With this, we get a rough idea of how long it takes to start working on a feature. When we start committing, we normally also have our first pipelines and tests in place. If we onboard lots of services, we could also measure the time between the first commit and the first successful pipeline run, as this can be a good indicator of the maturity of our templating mechanisms. From this point, it's always about making the feedback loop as short as possible but also as reliable as possible. Therefore, we are always measuring the duration of a pipeline run, but always track if the pipeline has been successful or failed. After some time, our feature is finished and delivered to an artifact store. This is the point where we could measure the time between the first commit and the first successful delivery. This can be a good indicator of the maturity of our CI and development process.

In the next steps, we might deploy our infrastructure and deploy this to the infrastructure. In this case, we could measure the time between the first commit and the first successful deployment. Furthermore, we could also measure the time between the delivery of the artifact and the deployment of the artifact in the first environment. This can be a good indicator of the maturity of our infrastructure and deployment process. Note that the **Infrastructure-as-Code** (**IaC**) and *deploy-as-code* processes repeat for each environment. Therefore, we could also measure the time between the delivery of the artifact and the first production deployment. As you see here, there are many things you can measure from a timing perspective, and you should decide which ones are important for your business and your customers.

In the upper part of the previous diagram, you can see some metrics we could measure in addition to the timing. In any case, you should always measure the success rate of your pipelines and deployments. As shown in the diagram, you can measure this for various process steps—for example, for the CI pipeline, but also for the IaC and the deployment steps. A lower success rate in one of the process steps can be a good indicator of a problem in this step. But it can also be a good indicator that this step works perfectly and that the problem is in the previous steps.

Previously, we talked about various things we could measure and where you can get creative. If you don't want to get creative, there are some metrics that can give you a good overview of the maturity of your deployment strategy. In *Chapter 1*, we discussed **DORA** metrics, which can describe the success of your deployment strategy. As a reminder, they are listed here:

- **Deployment Frequency** (**DF**): Frequency for delivering code to production
- **Mean Lead Time for Changes** (**MLT**): Time between commit and release
- **Change Failure Rate** (**CFR**): Changes that cause failures in production (as a percentage)
- **Mean Time to Recovery** (**MTTR**): Time to restore a system in the case of failure

In general, these metrics are based on velocity, but also on the error rate and recovery from errors. As failures can happen anytime, you should always have your **service-level objectives** (**SLOs**) in mind in addition. Therefore, a failure might be defined by some SLOs that have an impact on the CFR. When creating *safety nets* around your delivery process, it also makes sense to measure their effectiveness. Therefore, we should measure the deployment frequency not only at the pre-production stages but also how many failures are mitigated there. The same applies to tests. As these safety nets (as additional stages) reduce your velocity, you should be able to ensure that they are making sense.

Measuring such things is not as easy as it seems. You might use different tools for different parts of your delivery process. Therefore, you should always have a look at your tools and see if they can provide you with the metrics you need. If not, you should think about how you can get the data you need. In your build and delivery process, you might have some **unique identifiers** (**UIDs**) for your builds and deployments. These can be used to correlate the data from different sources. In addition, you should always think about how you can get the data you need. There are many tools out there that will help you on your journey. Nevertheless, it can also happen that you have to write some custom code to get the data you need. In the end, it's always about getting the data you need to make the right decisions.

In the next section, we will take a look at observability and how we can use it to measure the performance of our delivery strategy.

Observability

In the previous section, we discussed how we can measure the performance of our delivery strategy. In this section, we will clarify shortly what observability is and what we should know about it when starting to measure.

In fact, observability is the ability of a system to expose its internal state to the outside world, which means for us that we can measure the performance of our pipelines as well as deployment tooling. In the literature, observability consists of three pillars: logging, tracing, and metrics. We will now have a closer look at each of them, as follows:

- **Metrics** are numeric values (as we found in the previous section) that can be used to measure the behavior of a system. We can use metrics for dashboarding, but also for alerting. Normally, we know some types of metrics:

  - **Counters** are used to count things. For example, we can count the number of requests to our application. Normally, these metrics are always incrementing.

  - **Gauges** are used to measure the current state of a system. For example, we can measure the current CPU usage of our application. This type of metric can increment up or down.

  - **Histograms** are used to measure the distribution of a value. For example, we can measure the duration of a request to our application and get a histogram of the distribution of the duration.

  Typical tools for metrics are **Prometheus** and **Graphite**. In addition, many tools can be used to visualize metrics; the most prominent one in the open source world is Grafana.

- **Traces** are normally used to measure the path of a request through a distributed system. In a delivery scenario, we could also use traces for measuring the path of a request through our delivery system. As an example, we could start a trace when the build starts, add the context information when handing it over to the delivery system, and end the trace when the production environment is deployed.

  Typical tools for tracing are **Jaeger** and **Zipkin**. In addition, many tools can be used to visualize traces; the most prominent one in the open source world is Grafana.

- **Logging** is the process of collecting—more or less—unstructured data from the system and storing it in a central place. This data can be used for troubleshooting and is interesting in terms of delivery information when something goes wrong.

The power of observability is that we can combine the different pillars to get a better understanding of our system. For example, we can use metrics to measure the performance of our system but also use traces to see the path of a request through our system. In addition, we can use logs to get more information about the request. In the end, we can use all three pillars to get a better understanding of our system.

These basics should be enough to help you understand observability and how it can be used. When talking about deployment strategies in the next sections, we will need some observability to measure the performance of our delivery strategy.

Deployment strategies

In traditional systems, we often dealt with in-place upgrades or simple forms of rolling updates in which we replaced the whole application with a new version. In cloud systems, we have more options to deploy our applications but also have many possibilities to scale them. In this section, we will first have a look at scaling and autoscaling, and in the second step at the different deployment strategies. After that, we will make a short comparison between the different strategies. Finally, we will have a look at some tools that can help us with the implementation of these strategies.

Basics of scaling and deployment strategies

One of the basics of the various deployment strategies is scaling. Normally, we are distinguishing between two types of scaling—horizontal and vertical, as depicted in the following diagram:

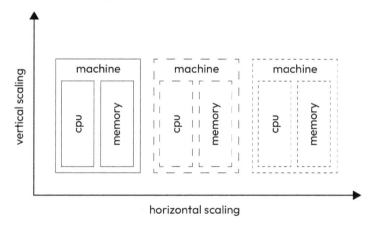

Figure 7.2 – Scaling

Let us look at them in detail:

- **Horizontal scaling**: When scaling horizontally, instances of the applications are added or removed. This is the most common type of scaling, as it allows us to scale workloads based on their load. It would also be possible to scale to zero when there is no load on the application. In this case, the application would be started when there is a request coming in. In serverless systems, this is very common. In other cloud environments, it's very common to scale down to at least one instance to keep the application running.

- **Vertical scaling**: To automatically change the dimensioning of the instances, we can use vertical scaling. This is often used to scale the memory or CPU of the instances.

In almost any strategy we will discuss here, we will start new instances of our application and then replace the old instances with new ones. Therefore, the capability of our systems to scale is instrumental for the success of our deployment strategies. We discussed the 12-factor app methodology in *Chapter 5*, so some of the points are already familiar to you. Let's have a look at the following points, which are important for the scaling of stateless workloads:

- **State**: The application should not store any state in a filesystem or in a local database. State information should be stored in a database or in a cache.

- **Data**: Avoid storing data in the local filesystem. If you have to store data, use external storage—for example, an object store such as **Amazon Simple Storage Service (Amazon S3)** or a database. At this point, it makes sense to think about building the database service on our own or using a managed service. Have in mind that it's not always easy to update databases in a rolling update, so if you are not comfortable with this, you should use a managed service.

- **Configuration**: Mount the configuration as a volume or use environment variables. This way, you can easily change the configuration without having to rebuild the image, which can help you configure the software for different environments.

In fact, there are a few questions you could ask yourself to check if your application is ready for scaling:

- What would happen if I started multiple instances of my application?
- What would happen if I deleted the local filesystem of my application?
- Can I change the configuration of my application without rebuilding the image?
- Is my backend capable of dealing with multiple frontends?
- What about the customer impact of all the things I just mentioned?

If all of these questions have no impact on your application and backends and if your customers are not affected by these changes, we're good to go and can scale our applications. Before we dive deeper into the deployment strategies, we will have a look at autoscaling. Autoscaling is a very important topic in cloud systems, as it allows us to scale our applications automatically and might help us to save money.

Autoscaling helps us scale our workloads automatically, based on metrics. For instance, we could define that the CPU consumption should not be higher than 80%, and the autoscaling mechanism will ensure that additional instances are started when this threshold is reached. We can use autoscaling based on at least CPU and memory consumption in many cloud services, mostly in virtual instances and in Kubernetes distributions. In many cases, it is also possible to use metrics from external providers—for example, to scale based on the number of requests to our application. In Kubernetes, we can implement such things as custom metrics, or with external tools, such as the **Kubernetes Event-Driven Autoscaling (KEDA)**.

At this point, we have all the basics we need to understand the different deployment strategies. In the following sections, we will have a look at the different strategies and how they work. At the end, we will make a short comparison between the different strategies.

Rolling updates

The first strategy we will have a look at is **rolling updates**, which is the most common and default strategy in Kubernetes. Using this strategy, a new instance is started, and when it's ready to receive traffic, the old instance is removed. In this way, we can update our application without downtime in a resource-friendly way. The following diagram shows the process of a rolling update:

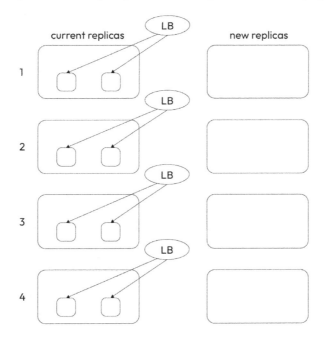

Figure 7.3 – Rolling update

Let's have a look at the preceding example, and see what happens:

1. All instances are running on the old (*current*) version of the application.

2. A new instance is started with the new version of the application, but traffic is still directed to all old instances.

3. When the new instance is ready to receive traffic, traffic is directed to the new instance.

4. The old instance is removed.

5. The process is repeated until all instances are updated.

As you can see, this can be implemented in a really resource-friendly way, as we only need the *desired number of instances + 1* instances to update all instances. Normally, this mechanism can also be configured to update multiple instances at the same time. In this case, we would need the *desired number of instances + number of instances to update at the same time* instances to update all instances.

Blue/green deployments

The second strategy we will have a look at is blue/green deployments. In fact, blue/green deployments are very similar to rolling updates, but we are using two different environments for the old and new versions of the application. The following diagram shows the process of a blue/green deployment:

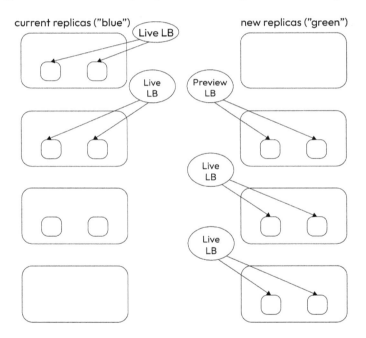

Figure 7.4 – Blue/green deployments

In this example, we are using two different environments, which are called **blue** and **green** in the first step, with all traffic directed to the blue environment. In the second step, we are starting new instances in the green environment, and we have two environments running in parallel. We can test the new version of the application in the green environment, and when we are ready, we can switch traffic to the green environment. In the final step, we can remove the blue environments.

This strategy is very useful when you want to run more sophisticated tests before switching traffic to a new version of the application. As an example, you can deploy a new version of the application to the green environment, run some functional or load tests to ensure that the new version fits into the working environment, and then switch traffic to the green environment. In this way, you can ensure that the new version of the application is working as expected, and you can switch traffic without any downtime.

Canary releases

Canary releases can combine the advantages of rolling updates and blue/green deployments. In this strategy, we are also using two different environments, but we are not running both environments in parallel. Rather, we are starting new instances in the green environment and directing a small percentage of traffic to the green environment. If the new version of the application is working as expected, we can increase the percentage of traffic directed to the green environment. In this way, we can ensure that the new version of the application is working as expected, and we can switch traffic without any downtime.

This strategy can be extended to multiple environments and could help you get more insights into the acceptance of new features or the behavior of the new version of the application. It can also be used to let users decide which version of an application they want to use. For instance, you could have three release channels—for example, *early access*, *stable*, and *long-term*—and let the users decide which version they want to use. In this way, you can ensure that the new version of the application is working as expected, and you can switch traffic without any downtime.

Feature flags

Feature flags are a very useful tool to enable or disable features in your application and can help you limit problems when delivering new features to your customers. In this strategy, you are instrumenting your application to enable or disable features based on a configuration. Using a configuration, you can enable or disable features in your application, and you can also define the percentage of users who should see the new feature. In this way, you can ensure that the new version of the application is working as expected, and you can switch traffic without any downtime. It also allows you to enable a feature and disable it again if you find any issues with the new feature.

Levels of deployment strategies

As you saw in the previous sections, the different strategies can help instill more confidence in the new version of the application. One thing to mention here is that we can use these strategies on different levels, as follows:

* **Environment level**: We can use these strategies to update whole environments—for example, we could have multiple production environments and update them one after another. We could also think of creating a new environment for each new version of the application, testing the application using canary or blue/green deployments, and removing the old environment after the new version is running in production.

- **Application level**: It's also possible to use these strategies on an application level—for example, we could have multiple applications running on the same environments and update them in a bundled way. This can be useful when workloads within an application are highly coupled and need to be updated together.

- **Workload level**: We can also use these strategies on a workload level, which would be the most granular. In this case, we can update every instance of an application individually. One thing to mention here is that this would be the most microservice-native way of updating applications, but is also the most challenging one for developers, as they need to ensure that the workload is capable of communicating with a variety of versions of its dependencies.

Depending on your preference and the applications you are running, you can choose the level of granularity you want to use, and you can apply various strategies to different levels. In the following section, we will have a look at how these strategies may impact your deployment process.

Things that change when you use a strategy

Especially when you are using Kubernetes, you might want to start with rolling updates to get your applications running in an easy way. Using liveness/health checks, you can ensure that your application is technically running and able to receive traffic. Perhaps these checks may not be enough to ensure that your application is really working as expected. For instance, your service might be able to receive traffic and the connection to any attached service is working, but the application gets terribly slow under load. In this case, you might want to proceed to a mechanism that gives you more confidence in the new version of the application before switching traffic.

Personally, I really like the idea of using blue/green deployments at the workload level, as it gives me the possibility to test the new version of the workload—even when deployed in parallel—before switching traffic. As an example, it's possible to deploy the application, find out how it behaves under some load, and then switch traffic. We will discuss this two-step process in more detail in the final section of this chapter.

When we are comfortable shifting traffic between multiple versions of the application, we can go for a more customer-centric approach and use canary releases. In this case, we will not only take care of the technical aspects of the deployed workload or applications but also the acceptance of the new version of the application by the users. Therefore, we rely heavily on customer feedback, and we can use this feedback to improve the new version of the application. Some of these things can also be put into metrics—for example, we can measure the acceptance of the new version based on business metrics or conversion rates. A special form of canary releases is the idea of using feature flags to enable new features for a subset of users. This can be a really powerful way to get feedback from your users and to improve the new version of the application.

Nevertheless, the usage of different deployment strategies can help us not only in supporting our business but also in enhancing our deployment process. In the following table, we can see some of the advantages and disadvantages of the different strategies:

Value	Rolling updates	Blue/green	Canary
Business	+ No downtime	+ No downtime + Preview environments	+ No downtime + Preview environments + Insights on user acceptance
Engineering	+ Small resource footprint + Easy to implement	+ Ability to test before switching traffic - High footprint	+ Shifting traffic gradually + Ability to react granularly to failures
Customer	+ No downtime - Possibility of unstable deployments	N/A	+ Optionally select the environment + Better usability - A bit hard to implement

Table 7.1 – Comparison of deployment strategies

The preceding table should give you a rough overview of the different strategies, their advantages, and their disadvantages. In any way you choose, you should always keep in mind that you should not only focus on the technical aspects of the deployment but also on the business value and the customer value. In the following section, we will have a short look at observability, which can help you to get more insights into your deployment process.

In fact, these strategies can help you avoid uncontrolled growth of your pre-production environments and help you keep environments clean and stable. On a small scale, this could limit the number of environments you need to run to two, which can help you keep costs low. Without stateful workloads, you might not even need two environments, as you can use blue/green deployments to test the new version of the application before switching traffic. Furthermore, you can use feature flags to enable/disable features in your application and therefore might have enough control over the new version of the application to operate in this way. In larger-scale environments, you will most probably have multiple environments, but the preceding mechanisms can help you keep the number of environments low.

In the next section, we will talk about the difference between deployments and releases.

Deploying and releasing

In the previous few sections, we have seen how we can use different strategies to update our applications. As discussed in the *Blue/green deployments* section, we can remove some of the complexity of the deployment process by using a two-step process. In this section, we will have a look at how we can split the deployment process into two steps—deployment and release. Let's look at these in a little more detail:

- **Deployment** is the process of taking code and deploying it to an environment. In this step, we are deploying the new version of the application and testing it in a controlled way.

- The **release** is the process of switching traffic from the old version of the application to the new version. In this step, we are switching traffic and making the new version of the application available to the users.

This two-step process can help us to get more insights into the deployment process and gain more confidence in the new version of the application. Some tools help us with this process, but they very often operate in the same way, as shown in the following diagram:

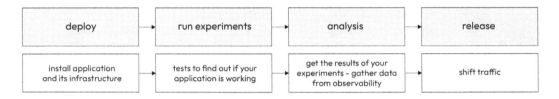

Figure 7.5 – Deployment and release process

In this diagram, we see that a deployment can start by deploying the software and its infrastructure. After that, we can run some tests to ensure that the application is working as expected. This could include the following:

- Simple tests to find out if the application is running

- Load tests to find out if the application is able to handle the expected load

- Functional tests to find out if the application is working as expected

- Performance tests to find out if the application is performing as expected

Additionally, we could also try out the application manually to find out if it's working as expected. In the third step of the deployment process, we analyze the results of the tests and can decide if we want to release the new version of the application. This is where our observability tools come into play. As an example, we could have defined that we want to release the new version of the application when the performance of the application is at least 90% of the performance of the old version. In this case, we could use our performance metrics to find out if the new version of the application is performing as expected. If the performance is not as expected, we could decide to roll back the deployment and try to fix the problem. If the performance is as expected, we can proceed to the release process. When the software is released, we will also have monitoring in place to find out if the new version of the application is performing as expected. If the performance is not as expected, we can take action to fix the problem.

By following such a process, we ensure that the software is running and performing as expected before we release it to the users. As deployment itself can be considered a critical process when software is upgraded in place, we take out some of the risks by deploying a new version of the application in a controlled way. In the cloud-native world, there are some tools that can help us with this process, as follows:

- **Argo Rollouts** is a progressive delivery operator that runs on top of Kubernetes. It helps us to manage the deployment process and to run different tests to ensure that the new version of the application is working as expected. It can also help us to release the new version of the application and to switch traffic between the old and the new version. Argo Rollouts is capable of doing canary and blue/green deployments and integrates well with many ingress controllers and service meshes, but also observability providers. It brings its own resource model, which is very similar to Kubernetes Deployments.

- **Flagger** is the progressive delivery controller for Flux. It can be used to manage the deployment process and to run different tests to ensure that the new version of the application is working as expected. It can also help us to release a new version of the application and to switch traffic between the old and new versions. Similar to Argo Rollouts, it can do canary and blue/green deployments and integrates with lots of tools. It uses Kubernetes primitives and adds its own resources to manage progressive delivery for them.

- **Keptn** helps you to manage the life cycle of Kubernetes applications and to automate the deployment process. It wraps around Kubernetes primitives to run analysis/evaluation before and after the deployment, but also adds the ability to install multi-service applications in a controlled way. For switching traffic between the old and new versions, it is possible to use Argo Rollouts or Flagger beside it.

In this section, we have seen how we can split up the deployment process into two steps: deployment and release. In the deployment step, we can run different tests to ensure that the new version of the application is working as expected. In the release step, we can switch traffic between the old and new versions.

Summary

In this chapter, we have learned why velocity is important in our deployment strategy and how we can use different strategies to update our applications. We have seen that we can use different strategies to update our applications, but we should always keep in mind that we should not only focus on the technical aspects of the deployment but also on business value and customer value. In the following chapter, we will have a look at security, how we can use them to run our deployment system securely, and how we can ensure that our deployment system is resilient to failures.

8

Security in Continuous Delivery and Testing Your Deployment

In the previous chapters, we discussed what we should consider when architecting our CI/CD infrastructure and why velocity is important. When deploying our software, we often deal with lots of moving parts and privileges, all of which can be abused. Furthermore, the build process itself might be vulnerable to attacks.

In ancient times, CI/CD environments were often treated as second-class citizens. This means that they were not secured properly and that they were often not maintained properly. This was mainly because they were not considered as important as the production environment.

However, this has changed recently. In the last few years, there have been many attacks on CI/CD environments. One of the most prominent attacks was the **Solarwinds hack**. The attackers compromised the CI/CD environment of Solarwinds and injected malicious code into their Orion Tool for updating IT infrastructure. This code was then distributed to Solarwinds customers and was used to compromise their infrastructure. This attack was one of the biggest attacks in recent years and was used to compromise many companies.

First, we will discuss some basics of security, such as zero-trust architecture and secret management. After that, we will inspect our pipeline and try to find some potential risks, as well as possible mitigation strategies. Finally, we will discuss how we can test our deployment and how we can deal with human error.

In this chapter, we will cover the following topics:

- Security basics
- Securing your CI/CD environment
- Making the pipeline secure

- Supporting tools and technologies

- Managing human error

Security basics

Security is important in every aspect of our lives, including our CI/CD environments. Just like you wouldn't leave your garage door open when you're not home, you shouldn't leave your CI/CD environment vulnerable to attacks. In this section, we'll discuss some basic security principles that you should follow when building your CI/CD environment. By following these principles, you can help protect your organization from attacks.

First, we should keep in mind that many CI systems can be considered **remote code execution as a service**. You are defining some tasks, scripts, and actions, and whatever these things are called in your tooling, and you execute them in an environment, hosted by you, or at least under your responsibility. In many cases, this environment has access to your code repository or other internal systems, which could be very interesting for attackers. As we learned earlier in this chapter, there were **supply chain attacks** that compromised lots of customer systems. As in every other system, one of the first steps we should take is to define our business goals for securing the system:

- Our customers should be able to trust us and the software we are delivering

- We should minimize the risk of attacks on our infrastructure

- If an attack happens, we should be able to detect it as soon as possible and react accordingly

These are just simple examples, but they should give you an idea of where this journey should go. As a short recap of common security principles, there is the **CIA triad**, which stands for **confidentiality, integrity, and availability**. This triad is often used to describe the goals of security:

- **Confidentiality**: This means that only authorized users should have access to the data. For our CI/CD environment, this means that only authorized users should have access to it and that only authorized users should be able to change the configuration of our CI/CD environment.

- **Integrity**: This means that we should make sure that our build artifacts are not modified by unauthorized users. As a result, we should make sure that our build artifacts are signed and that we can verify the signature.

- **Availability**: This means that our CI/CD environment should be available when we need it. This means that we should make sure that we have redundancy in place and that we can still operate our CI/CD environment if one component fails.

To achieve these goals, we should follow some basic security principles. One of the very first ones related to confidentiality is the **principle of least privilege**. This means that users should only have the privileges they need to do their job. This means that we should make sure that users do not have more privileges than they need. This is especially important for our CI/CD environment as it is often used to deploy our application to production. As a result, we should make sure that only authorized users can deploy to production. This can be achieved by using a central authentication mechanism, such as **lightweight directory access protocol (LDAP)** or **Active Directory**. This makes it easier to manage users and their privileges, especially when they are leaving the company. This is also valid in today's more and more source-control-driven world, as we will discuss later in this chapter.

Furthermore, we should make sure that we have multiple layers of security in place. This is also known as **defense in depth**. If someone can bypass one security mechanism, the attacker should not be able to bypass all of the other security measures. As an example, if an attacker can compromise our CI/CD environment, they should not be able to deploy to production. Another additional layer of security would be network segmentation and policies, which only allow read access to the source control management system from the CI environment and to the target systems it needs to create the deployment artifacts. This makes it harder for an attacker to compromise our infrastructure as they need to compromise multiple systems to be able to deploy to production.

Another modern term we hear today is **zero-trust architecture** (as described in *NIST.SP.800-207*), which moves away from the traditional network-based approach to users, assets, and resources (*NIST, 2020*). One of the major points of the zero-trust architecture is the fact that security should not be defined based on the location of a system, but on the identity of the user. This means that we should not trust a system just because it is in our network, but we should verify the identity of the user. This is especially important for our CI/CD environment as it is often used to deploy our application to production. As a result, we should make sure that only authorized users can deploy to production.

We could write a whole book about security principles and how to apply them to our CI/CD environment. However, this is not the goal of this book. The goal of this book is to give you an idea of what you should think about when you are building your CI/CD environment. As a result, we will now discuss some of the most important aspects of securing your CI/CD environment, based on a real-world example.

Securing your CI/CD environment

Let's assume that we want to deliver some application our company is maintaining and that as a modern company, we are using CI/CD in a cloud environment, as shown in the following figure:

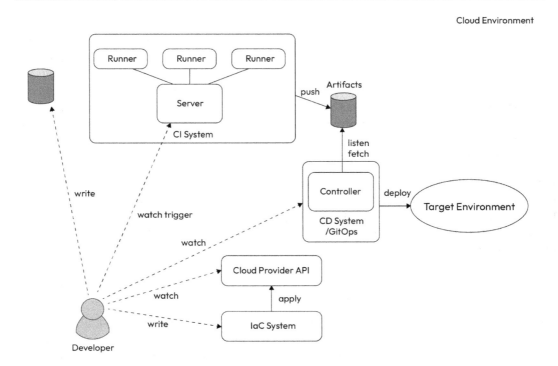

Figure 8.1 – Example of a CI/CD environment

The preceding figure shows a simplified environment that is hosted in the cloud and was created using IaC. We assume that application development and deployment processes are mainly source-code-driven and that we are using a **source code management** (SCM) system, such as Git. We have a CI system that listens to changes; when something changes, it builds and tests software and pushes this to an artifact store. We have a GitOps controller in place, which listens for changes, and when new artifacts arrive or the repository gets updated, it deploys the application. This is a very simplified process for the sake of finding some attack vectors. In a real-world scenario, this might not fit into one page and more components might be involved. However, this is not important for our discussion. From an infrastructure perspective, let's find some attack vectors:

- **IaC system**: Someone could inject malicious code into the IaC system, which could lead to a compromise of the whole infrastructure. if someone gains access to the IaC system, the attacker could deploy their own infrastructure there and use it for their own purposes. This could be used to mine cryptocurrency or host malicious content. Furthermore, the attacker could also use the IaC system to deploy malicious infrastructure, which could be used to attack other systems. As a result, we should make sure that only authorized users can access the IaC system and that we have a way to detect if someone is trying to inject malicious code into it.

- **SCM system**: Very often, pipelines/workflows are defined in the SCM system and the CI system listens for changes in the SCM system. The CI system often simply executes the code that is stored in the SCM system. Therefore, we can assume that if someone gains access to the SCM system and injects malicious code, the CI system will execute it. As a result, we should make sure that only authorized users can access the SCM system and that we have mechanisms in place (**pull requests**) to detect if someone is trying to inject malicious code into the SCM system.

- **CI system**: The CI system is used to build and test the application and provides **remote code execution as a service**. This means that we can execute arbitrary code on the CI system. To mitigate some risks there, we could limit the privilege of the CI system and make sure that it can only access the resources and execute the commands it needs to build and test the application. Furthermore, we should make sure that only authorized users can access the CI system.

- **Artifact store**: The artifact store is used to store the deployment artifacts. When someone gains access to the artifact store, artifacts can be changed and the attacker could inject malicious code into the artifact store. A good practice is to grant access to artifact stores based on the environments and manage the artifact store's life cycle through a promotion mechanism. In this way, for instance, production systems can only pull artifacts that have been promoted (signed) to be released in the corresponding stage. Therefore, in a perfect world, only the CI system has write access to the artifact store. Furthermore, there should always be a way to verify the integrity of the artifacts (signatures).

- **GitOps controller**: The GitOps controller is used to deploy the application. As this controller often has wide privileges on the target system, it would be very easy to deploy malicious applications to the target system. Therefore, we should make sure that only authorized users can access the GitOps controller and that we have a way to detect if someone is trying to inject malicious code there.

- **Target system**: The target system is the system where the application is deployed to. If someone gains access to the target system, they could deploy malicious applications there. It's also possible that someone compromises the GitOps controller and tries to deploy malicious applications to the target system. Therefore, we could have policy enforcement controllers in place that make sure that only authorized applications can be deployed to the target system. This is always a good idea if there is no static write access for humans on the target system.

As you can see, there are lots of attack vectors in a CI/CD environment and lots of things we could consider when building a CI/CD environment. However, we should always keep in mind that security is a process and not a product. Therefore, we should always try to improve our CI/CD environment and make it more secure. To get started, let's build a short checklist of things we should consider when building a CI/CD environment:

- **General**:

  - Is a central authentication mechanism in place?

  - Is multi-factor authentication enabled?

- **Securing IaC system/cloud provider API**:

  - Who can access the IaC system/check in code?

  - Are policy checks in place?

  - Which privileges does the IaC system have?

- **Securing source code repositories**:

  - Can only authorized users write access to the repository?

  - Are peer reviews (pull requests) and branch protection rules in place?

- **Securing the CI system**:

  - Who can access the CI system?

  - Is the CI system running in a separate network?

  - Is the CI system running in a separate account?

  - What privileges does the CI system have?

  - Which commands can be executed on the CI system?

- **Securing the artifact store**:

  - Who has access to the artifact store?

  - Can tags be overwritten (immutable tags)?

  - Are there any integrity checks in place?

  - Are my artifacts signed?

  - Are my artifacts checked for vulnerabilities?

- **Securing the GitOps controller**:

  - Who has access to the GitOps controller?

  - Can the configuration on the GitOps controller be changed manually?

  - Is auditing in place?

  - Which repositories can be deployed?

- **Securing the target system**:

  - Is a policy enforcement controller in place?

  - Are there **role-based access control (RBAC)** rules in place?

  - Is auditing in place?

Many other things might be relevant to your system, so feel free to add them to the list. However, this list should give you a good starting point for building a secure CI/CD infrastructure. Before we take a look at the tools that might help us build a secure CI/CD infrastructure, let's take a look at how we can secure our pipelines.

Making the pipelines secure

Pipelines/workflows are a fundamental part of a CI/CD infrastructure. As we discussed earlier, we must know what is running and what we're building. To proceed with our example, let's take a look at the following pipeline:

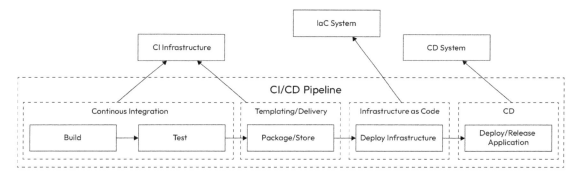

Figure 8.2 – Example pipeline

This is kind of a different view of the system we've seen before, but this one is more focused on the pipeline. As you can see, we have the following building blocks:

- **Continuous integration and templating**: In this step, we are building our application and creating the deployment artifacts. Furthermore, we are creating the infrastructure templates that are used to deploy our application.

- **IaC**: In this step, we are deploying our infrastructure using the infrastructure templates.

- **Continuous delivery**: In this step, we are deploying our application to the infrastructure that we created in the previous step.

In each of these steps, different tools can be involved and there are many things we should consider when building a secure pipeline. Let's take a look at the different steps and what we should consider there.

In general, we should always take care of the things we're running. In some tools, there are templated actions or libraries that we could use. As an example, there might be a predefined action that helps us build a Docker image. In any case, we should always make sure that we can trust this action/library. Therefore, we might want to check the trustworthiness of the action/library and maybe even check the source code. Furthermore, we should always make sure that we're using the latest version of the action/

library. In some cases, there might be vulnerabilities in the action/library that can be exploited by an attacker. Therefore, we should always make sure that we're using the latest version of the action/library.

Additionally, we should always make sure that we're using the latest version of the tools we're using. In some cases, there might be vulnerabilities in the tools that could be exploited by an attacker. Therefore, we should always make sure that we're using the latest version of the tools. There are tools out there such as **Dependabot** and **Renovate** that help us update our dependencies, as well as our CI/CD infrastructure.

At some point in our pipeline, we might want to test the things we built, maybe even in a production-like environment. Depending on the type of application we're building, we might want to test the application in a container or a virtual machine. In some cases, this might not work because the application can't run in a container or virtual machine. In this scenario, we might want to test the application on a real system. Even in these cases, treat the system as a production system and make sure that you have the same security measures in place.

Especially in open source environments, you should be aware of who can contribute directly to your repository. You might have secrets in your repository, which could be read or printed out by an attacker in a changed pipeline configuration. Furthermore, a user might push a malicious commit to your repository. Therefore, you should always make sure that you're using branch protection rules and peer reviews (pull requests) for your repository. You could avoid many of these issues by using a fork-based workflow. In this case, you're not able to push directly to the repository, but you have to create a pull request. If an attacker pushes something to their fork, you're still able to review the changes before merging them into your repository.

In the previous few sections, we talked about the things that could happen due to an unsecured CI/CD infrastructure and pipelines. In the next section, we will look at the tools that might help us build a secure CI/CD infrastructure.

Supporting tools and technologies

In this section, we will take a look at the tools and technologies that might help us build a secure CI/CD infrastructure. We will take a look at the following tools and technologies:

- Central authentication
- Secrets management
- Policy enforcement controllers
- Auditing

Given the things we discussed previously, these tools and technologies might help us avoid some of the issues we discussed. However, they are not silver bullets and you should always make sure that you're using them in the right way. Let's take a look at the different tools and technologies.

Central authentication/identity management

In the example we used in *Figure 8.1*, we are dependent on lots of components, all of which might need all the necessary credentials. Think of a scenario where the password of one user is compromised and you have to change it, or a user leaves the company. Each of these cases causes lots of work and with a rising amount of used systems, this might become a nightmare. Therefore, we should always try to use a central authentication system that is used by all systems.

With the rise of the cloud, there are many things out there that can help you keep your identities consistent across clouds, as well as on-premises. As an example, you could use Azure AD to manage your identities in the cloud and your data center. Furthermore, you could use Azure AD to manage your identities in different clouds, such as **AWS**. In this case, you would have a central place where you can manage your identities and you would not have to worry about the different systems. Another great benefit of using a central authentication system is that your company has a central place where you can manage your identities. This means that you can easily add or remove users from your company, and you don't have to worry about the different systems.

There are also similar services out there, such as Google Cloud Identity, Okta, and Keycloak on-premises. In the end, it doesn't matter which service you're using, but you should always make sure that you're using a trustworthy central authentication system.

Secrets management

One of the major threats in CI/CD environments is keeping secrets secret and not exposing them somewhere. In our example, we have lots of secrets, all of which are used in different places. As an example, we have the following secrets:

- Cloud account credentials (IaC)
- Repository credentials (continuous integration)
- Artifact repository credentials (continuous integration and continuous delivery)
- Credentials for the target system (continuous delivery)

As you can see, there are lots of secrets that are used in different places and can point to different systems. While we don't care about the functionality of the secrets, we must ensure we store them securely:

- **Storing secrets in Git**: You might be wondering why we should store in Git. In GitOps-driven workflows, it might be a good idea to deliver them alongside the rest of the code and use them on the target infrastructure. Nevertheless, you should never store secrets in plain text in Git so that there are ways to encrypt them. There are tools out there such as **Mozilla SOPS** and **Bitnami Sealed Secrets** that can assist you in encrypting and decrypting your secrets. Always keep in mind that you also might need a way to store the key for encrypting and decrypting your secrets. This could be achieved by using another key management system at your cloud provider.

- **Storing secrets in your cloud provider**: Many cloud providers offer a way to store secrets securely. As an example, you could use Amazon **Key Management Service (KMS)** or Azure Key Vault to store your secrets. In this case, you would have a central place where you can store your secrets and you don't have to worry about the different systems. Furthermore, you can also use these services to encrypt and decrypt your secrets. These services can also be integrated with platforms such as Kubernetes. In this case, you don't store the secrets in Kubernetes itself; instead, you store them in the KMS and Kubernetes will only have access to the secrets. This is a great way to store your secrets securely and you don't have to worry about the different systems.

- **Storing secrets in your CI system**: You will need some secrets to access your systems in the CI system. Therefore, many CI systems offer a way to store secrets in a secure way, which is comfortable to some extent. Often, these secrets might not be accessible from external systems, so keep in mind that you might need to store them in another system as well. Sometimes, CI systems also offer a way to access external systems, such as AWS, Azure, or GCP. In this case, you might be able to use the secrets from the CI system to access the external systems.

- **Self-hosted secrets management**: There are also tools out there that can help you store your secrets securely. As an example, you could use **HashiCorp Vault** for this. In this case, you would have a central place where you can store your secrets and you don't have to worry about the different systems.

There are many other secret management tools out there (such as **Bitwarden**, **1Password**, and others) that you can use in your CI/CD environment. In the end, it doesn't matter which service you're using; just make sure that you're trusting the service and that you're using it in the right way.

Policy enforcement controllers

Since we talked about Defense in Depth and different layers of security, we should also talk about **policy enforcement controllers**. Normally, we should try to avoid non-programmatic access to the production environment. But even when this is implemented, the GitOps controller might be compromised. Therefore, policy enforcement controllers, such as **Kyverno** or **Open Policy Agent**, might help us avoid unwanted changes in our production environment. Something we could enforce, especially in Kubernetes environments, is **limiting container registries**, which can help us avoid someone running a container from a malicious registry. It would limit the usage of privileged containers, thus avoiding someone running a container with root privileges. It would also limit the usage of hostPath volumes, thus avoiding someone mounting a volume from the host system. Enforcing resource limits would help us stop someone from running a container that is using too many resources. Ensuring signed images would also help us stop someone from running a container that does not originate from a trusted source.

One thing to consider in this case is that the policies should not be applied by the same system that is deploying the application. Otherwise, the policy enforcement controller might be compromised as well. Therefore, you should use a different system to apply these policies. As an example, you could apply the policies with your IaC tool or even with another GitOps controller. In this case, you would have a second layer of defense that would help you avoid someone compromising your production environment.

Auditing

Last but not least, we should talk about auditing. **Auditing** is a very important part of security and should be done in every environment. In this case, we should also audit our CI/CD environment. As an example, we should audit the following things:

- Who has access to the CI/CD components?

- Who logged in when and on which system?

- What was done on the system?

- What was changed on the system?

Such things are highly dependent on the software you are using. Many tools provide a way to audit the system, write logs, or send events to a system dealing with them. Make sure that you have a way to analyze logs if you are auditing your CI/CD environment. Otherwise, you might not be able to find out what happened in your CI/CD environment.

These things can be achieved by observability solutions such as **Loki** or **Elasticsearch**. In any way, you can make sure that your logs are not only stored on the system you are using but also in a central place. Otherwise, an attacker might be able to delete the logs on the system and you won't be able to find out what happened.

In this section, we talked about some tools that can assist you in securing your CI/CD environment. In the final section of this chapter, we will talk about human error, why this is important, and how to learn from it.

Managing human error

Even though we have lots of tools and measures to secure our environment, we might not be able to avoid human error. As an example, the following things might happen:

- Someone might accidentally delete a production environment on the cloud environment

- A developer might accidentally check in secrets

- A network administrator might accidentally open a port to the public

These things might happen and we should be prepared for them. In the next few sections, we will talk about how to deal with human error and how to learn from it.

Building a culture that allows for mistakes

The first thing we should do is build a culture that allows for mistakes. This might sound strange, but it is very important. If you are working in a culture where mistakes are not allowed, people might try to hide their mistakes. This might lead to even bigger problems. Therefore, we should build a culture where mistakes are allowed and where people can talk about their mistakes. This will help us to learn from our mistakes and avoid them in the future.

Therefore, we should apply the following rules:

- Allow people to make mistakes (*"Why didn't you break something today?"*).

- If someone makes a mistake, help them fix it (*"I could be the one who makes the next mistake."*).

- When something happens, it is not the fault of the person who made the mistake; instead, it is the fault of the system that allowed the mistake to happen (*"We should find out why this mistake was possible and how we can avoid it in the future."*).

When something goes wrong, we should not blame the person who made the mistake (unintentionally); instead, we should try to find out what went wrong and how we can avoid it in the future. As an example, if someone accidentally deleted a production environment, it could be a very good idea to rethink why this person had access to the production environment. In the next step, we might find out that the write access to this environment was not even necessary and that we should avoid this in the future. In this case, we learned from the mistake, and we can avoid it in the future.

Talking about errors

In any case, we should avoid other people being able to run into the same problem. Therefore, we should talk about our errors and mistakes in public. As an example, some meetups have sessions where only production errors are discussed and how they can be avoided. This is a very good way to learn from mistakes and avoid them in the future.

Post-mortems/documentation

You might have experienced that an application you needed broke and you spent some time fixing it. If you were the customer of the application, you might have wondered when the application would run again and why it broke. Therefore, it is very important to document things, tell people why the system is not working, and keep them up to date. The more important thing is that you should document what went wrong and how you fixed it.

Let's consider that one of your developers accidentally deployed the wrong version of the application. First, the developer should inform the team of this, as well as the customers. The team will help them fix the issue. Furthermore, it would be a good idea to document the steps to resolve the problem. After the initial state has been restored, there should be a **post-mortem meeting** to discuss what went wrong, why it went wrong, and how to avoid it in the future. To make the situation a bit more comfortable, we could also write down what went well during the process and where we were lucky. In any case, the whole process should be documented and the documentation should be shared with the team and the customers. This helps us avoid making the same mistake in the future and learn from it.

Automation/testing

At the end of each post-mortem, the answer to most problems will end up in automation or involve adding more testing. Going back to the network administrator who accidentally opened a port to the public, we might want to automate such processes and add more reviews to it. As an example, we could add a review process to the firewall rules so that someone else has to approve the changes. Furthermore, we could automate the process of opening a port to the public so that the network administrator does not have to do it manually. In this case, we would have automated the process and added more testing to it. This would help us avoid making the same mistake in the future.

In today's world, we should try to automate as much as possible to avoid human error. This can also happen incrementally, as follows:

- **Manual process**: One person is doing the whole process manually; they have the necessary knowledge and they are the only one who can do it.
- **Manual documented process**: One person is doing the whole process manually; they have the necessary knowledge and they have documented the process. Therefore, they enable other people to do the same process.
- **Semi-automated process**: The steps needed to fulfill the process are automated (even partially) and triggered manually. Therefore, the process is faster, can be done by other people, and is less error-prone.
- **Fully automated process**: All steps are automated and triggered automatically. Therefore, the process is fast, can be done by other people, and is less error-prone.
- **Continuous improvement**: When the process is automated, we can start to improve it. As an example, we can add more testing to it, add more reviews to it, add more automation to it, and so on.

The major aspect you should take away from this section is that everyone can make mistakes, but we should learn from them and try to avoid them in the future. With a good culture, we can avoid many mistakes and with a high grade of automation, we can avoid even more mistakes and improve our processes continuously.

Summary

In this chapter, we dealt with the security of our CI/CD environment. We talked about the different components of a CI/CD environment and how to secure them. Furthermore, we talked about how to secure the CI/CD pipeline itself and how to secure the CI/CD environment. In the end, we talked about human error and how to deal with it. In the next chapter, we will talk about how to secure our applications.

Part 3:
Best Practices
and the Way Ahead

In this part, we will describe some best practices and future trends in continuous delivery. Furthermore, you will get an idea about the open source ecosystem and how to be a good citizen in the open source world. Finally, we will have a knowledge assessment to find out how much has been learned throughout the book.

This part contains the following chapters:

- *Chapter 9, Best Practices and References*
- *Chapter 10, Future Trends of Continuous Delivery*
- *Chapter 11, Contributing to the Open Source Ecosystem*
- *Chapter 12, Practical Assignments*

9

Best Practices and References

In this chapter, we focus on the optimal posture for continuous delivery in the cloud to maximize business efficiency. Often, best practices are the result of tasks, processes, and tools implemented by various teams over a period of time that led to them experiencing substantial increases in efficiency. Although we discussed many of the best practices in the previous chapters, we will summarize them here and also provide references to them.

This chapter covers the following topics:

- Establishing and adopting best practices
- End-to-end responsibility
- Developing in a cloud-friendly way
- Delivery best practices
- Security best practices
- More best practices
- Case studies

Often, best practices are adopted to achieve specific business outcomes. In some cases, best practices can also be contractually binding and can even be required by regulatory bodies. Before we start to describe the best practices, it is important to understand how best practices are developed and where we can implement best practices.

Establishing and adopting best practices

When building a new delivery system, you should be aware that you are not the first one to do so. Teams are building new delivery systems every day and learn from their mistakes and successes. Often, some processes and tools they use are adopted by many other teams and people, and after a while, they may emerge as best practices. However, although certain best practices might work for many teams, they might not work for you. Therefore, it is important to understand the context in which the best practices were developed and how they can be adopted in your context. Even the best practices we describe in this book might not be the *silver bullet* for your team and adopting all of them could be a real challenge. Therefore, always assess best practices carefully in your context and adopt them incrementally. In some cases, you might even decide to not adopt a best practice at all.

Before we go more into detail about specific best practices, we will discuss the different areas in which best practices can be applied. This will help you to understand the context in which the best practices were developed and how they can be adopted in your context:

- **Internal best practices**: Often, internal best practices are established to solve a specific business problem, such as optimizing a flow or improving resource efficiency. Organizations develop these best practices through trial and error. These best practices are often not published and are not available to the public. However, they are often shared with other teams within the organization.

- **Functional best practices**: Often, functional best practices are established through specific organizations working to establish functional standards and best practices. An example of such organizations is the **National Institute of Standard and Technology (NIST)**, which is very active in outlining best practices in the domains of DevSecOps and cybersecurity. Another example is the **Institute for Standardization and Organization (ISO)**, which actively works with standards and best practices. When it comes to open source tools and technology, the **Cloud Native Computing Foundation (CNCF)** and the **Continuous Delivery Foundation (CDF)** are very active in the cloud-native functional area.

- **Industry benchmarking**: Industry benchmarking is important as a key reference point and provides a measurable criterion for evaluating the posture of the organization. Examples of such benchmarks include the **DevOps Research and Assessment (DORA)** report and the State of Continuous Delivery report by the CDF, and – last but not least – **Supply-chain Levels for Secure Artifacts (SLSA)**.

In the following sections, we will describe best practices in different areas of continuous delivery in the cloud that we and other experts have found useful.

End-to-end responsibility

If you are familiar with the DevOps movement, you might have heard the phrase, *"You build it, you run it,"* from Werner Vogels. This means that the team that builds a service is also responsible for running it and even if there are other teams involved, developers should behave as if they will be the ones on-call for this service. Therefore, the operational aspects of a service have to be considered from the beginning and operating the service should be made as easy as possible. This is also one of the considerations under *Operational Excellence* in the **AWS Well-Architected Framework**.

This responsibility extends across all of the building blocks of a service, including the infrastructure, the application, and the delivery pipeline, but also security. Building silos should be avoided and teams should be empowered to take responsibility for their services. This also means that teams should be able to make decisions about the tools and processes they use to build and operate their services. This is also one of the reasons why we recommend using a self-service approach for building delivery systems.

Furthermore, **observability** is one of your first-class citizens when building a new service, but also the delivery system itself. As we have mentioned throughout this book, you must have a clear understanding of the health of your delivery process and the services you are building. Therefore, you should implement observability from the beginning and make sure that you have the right metrics in place to understand the health of your delivery system and the services being built. Using the DORA metrics is a good starting point, but having a distributed trace for your delivery system can help you find bottlenecks and issues in your delivery system.

Developing in a cloud-friendly way

It's always good to have a clear understanding of the cloud provider and platform you are using. This involves the services you are using as well as the delivery and service model. Although development in the cloud is not really the main focus of this book, we will discuss a few best practices that have a direct impact on the delivery process.

Loose coupling

When building cloud-native applications, you should always aim for **loose coupling**. This means that you should avoid tight dependencies between your services and should use asynchronous communication between your services. Doing this will make it easier to scale your services independently. Furthermore, it will also make it easier to replace and deploy new versions of your services. This also means that you should avoid using shared databases, opting for event-driven architectures instead. This also avoids inter-service dependencies and simplifies your deployment process.

12-factor apps

In *Chapter 5*, we introduced the **12-factor app methodology** briefly. When developing applications for the cloud and especially on container platforms, these 12 factors are a good starting point and you should get familiar with them. Although the 12-factor app methodology is not a silver bullet, it can help you to build cloud-native applications that are easy to deploy and operate. We have summarized a few of the 12 factors important for the delivery process as follows:

- **Build, release, run**: Separate your build, release, and run stages. This makes it easy to build and release your application and it also makes it easy to run your application in different environments. This also means that you should not build your application during the deployment process, but rather beforehand and store it in a registry. Having a pre-built artifact ensures that you can deploy the same artifact to different environments and roll back to a previous version. Furthermore, it also makes it easier to run your application locally and to test it.

- **Logs**: Logs are a first-class citizen in cloud-native applications. Therefore, make sure that you are logging the right things and that you have a proper log management system in place. Treat logs as event streams and make sure that you don't log to files, but to `stdout/stderr`. This will make it easier to integrate your application with other tools and services. Furthermore, make sure that you are not logging sensitive information and that you are using structured logging.

We also discussed config and dev/prod parity in *Chapter 5*, which might be the most important factors for the delivery process. Especially when you are starting your journey to the cloud, you should make sure that you are at least familiar with the 12 factors.

Use proper versioning, avoid "latest"

Throughout our whole delivery process, we are highly dependent on the versioning of our artifacts. Therefore, we should ensure that each artifact has a proper versioning scheme we could use in later steps of our delivery workflow. Additionally, we also want to ensure that our builds and deployments are reproducible. Therefore, we should avoid using `latest` tags for our artifacts and use proper versioning schemes instead.

There are different levels of versioning we should consider when building a delivery system:

- **Artifact versioning**: The version of each artifact we are building
- **Service versioning**: The version of each service we are building
- **Platform versioning**: The version of the platform we are using

To make this a bit clearer, let's have a look at the following example:

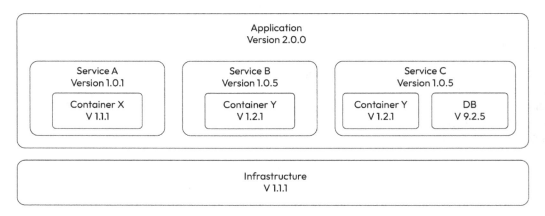

Figure 9.1 – Levels of versioning

This example shows the various levels of versioning we have to consider when building applications. You could have different services bundled together in one application. This application can also have infrastructure constraints and therefore it can be very useful to bind an application deployment to a specific version of the infrastructure. The most common versioning scheme is **semantic versioning**, but you can also use other versioning schemes. The important thing is that you have a clear understanding of the versioning scheme you are using and that you use it consistently across all of your services.

Split configuration in reusable parts

As we do while developing software, writing IaC, and pipeline configurations, GitOps configuration should usually be split into reusable parts (as long as it makes sense and provides benefits).

As an example, there can be a foundational base that provides a minimum usable construct (and may include company-specific policies also). Any service or application building on top of the base may override or extend parts. The base itself is already referenced using versioning. When releasing new versions of the base configuration, all dependencies just need to point to the new version. Using such a split setup provides better security by restricting access. It may even provide more flexibility (as an example, providing more permissions to devs on non-production environments while restricting any access to production completely and only allowing changes using promotions). Furthermore, it may provide better reusability and consistency.

In **Infrastructure as Code (IaC)**, such modularized setups can have a big impact on your system. Think of a configuration where each commit to the main branch of a repository triggers a pipeline that applies the configuration to the target system. If you have a monolithic configuration without proper versioning, you may end up in a situation where you apply a change to a non-production environment that breaks the production environment. If you have a modularized setup, you can apply changes to non-production environments and test them before applying them to production. Furthermore, you can also apply changes to production environments in a controlled way, for example, by using a **canary deployment**. This can be done by using a modularized setup and proper versioning as shown in the following example:

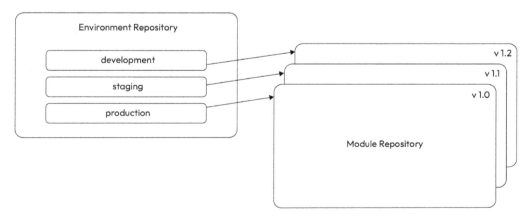

Figure 9.2 – Modularized configuration

This example shows that you can store the configuration of your environment in one repository (**Environment Repository**), and only put references to the modules in this configuration. When you consider a change to be stable, you can update the reference in the next environment and apply the change to the next environment. There are similar mechanisms for other configuration types, for example, this can easily be adapted with pipeline configurations (e.g., **Jenkins Shared Libraries**) or GitOps configurations (e.g., chart/repository versioning).

This section leads nicely onto the next section, where we will discuss some best practices for the delivery process.

Delivery best practices

In this section, we will get to the core theme of this book: the delivery process itself. We will discuss some best practices that can help you to improve your delivery process and build a process suitable for your needs.

Pipeline as Code

Throughout this book, we discussed some examples of codified configuration such as IaC and Declarative Configuration in GitOps. This approach can and should also be applied to the delivery process itself. This means that you should avoid configuring your delivery process in a UI (**ClickOps**) and instead configure it in a declarative way. This provides the big two advantages that your delivery process is versioned and that you can easily reproduce it. Furthermore, it also makes it easier to test your delivery process and run it locally, along with finding and fixing bugs.

Treat every build as a release candidate

In the past, it was common to have a pipeline that built some artifacts that were considered for testing and when everyone attested this version as stable, a separate release build with a different version was created. Therefore, it sometimes happened that during a release cycle, the code base in the main branch of a repository was not very stable and the release build was not the same as the build that had been tested. Furthermore, this approach could also have a negative impact on the velocity of the team, because the team had to wait for the release build to be created before they could test the build.

When aiming for velocity, every build should be considered a release candidate, and, in a perfect world, should be deployed to production if it passes all tests. This approach has several advantages:

- Features go to market very fast
- The awareness of the team regarding quality is increased
- The team is more aware of the impact of its changes
- The testing and release processes are more stable because they are used more often

As with every approach we are discussing in this book, we are not saying that you should do this in every case. However, at least treating every build as a release candidate is a good approach to improve your delivery process.

Automated testing and quality gates

Hand in hand with the previous section, automated testing is a very important part of the delivery process. When doing continuous delivery, you should aim for high test coverage and should try to automate as many tests as possible. This will help you to ensure that your application is working as expected and that you can deploy your application to production without any manual intervention.

These tests, together with quality gates, are very handy when it comes to the decision of whether a build should be deployed to production or not. As an example, you could build and deploy your application to the test environment, run some functional and performance tests, and find out that the performance of your application is not as expected. In this case, you could automatically stop the deployment to production and inform the team about the issue. This will help you to protect against deploying any faulty code to production and avoid negatively impacting your customers. On the other hand, you could also use quality gates to automatically deploy your application to production if all tests are passed. This will help you to ensure that you are deploying your application to production as fast as possible.

This can be enhanced by using blue-green or canary deployments as we described in *Chapter 6*. This will help you to ensure that you do not negatively impact your customers when deploying a new version of your application to production and that you can easily roll back to the previous version if something goes wrong.

Preview environments

Even if you do use automated testing and quality gates, you may still want to test the behavior of your application in a production-like environment, for example for acceptance testing. In this case, you can use preview environments. **Preview environments** are environments created on demand and destroyed after a certain period of time. They are usually created for a specific pull request and they are used to test the behavior of the application in a production-like environment. This can be very handy when it comes to acceptance testing, because you can test the behavior of your application in a production-like environment without impacting your customers.

These preview environments can get really large and complex, so automation is key here. You should aim for the fully automated creation and destruction of these environments. This will help you to ensure that you are not wasting any resources nor creating any security issues. Furthermore, you should also aim for the fully automated deployment of your application to these environments. This will help you avoid wasting any time, and allow you to easily test the behavior of your application in a production-like environment and reproduce your environment when you need to debug an issue.

In this section, we discussed some best practices for the delivery process. In the next section, we will discuss some best practices we found useful when it comes to securing delivery processes.

Security best practices

We discussed security-related topics in the previous chapter. In this section, we will add some more best practices that we have found useful when it comes to securing delivery processes.

Supply-c Levels for Software Artifacts (SLSA)

SLSA (pronounced salsa) is a security framework of best practices for ensuring the integrity of software artifacts throughout the entire software supply chain. It consists of incrementally adopted security guidelines for the software supply chain, offering a four-level hierarchy of maturity, where the fourth level is the desired end state:

1. **SLSA 1**: Indicates adoption of fully scripted/automated build processes and generate provenance statements. They display evidence of how the artifact was built, including the build process and the dependencies.

2. **SLSA 2**: Indicates adoption of version-controlled deliveries and a hosted build service that generates provenance.

3. **SLSA 3**: Indicates adoption of source and that the built platform meets the standards to provide auditability of the source and provenance.

4. **SLSA 4**: Indicates adoption of a two-person review process for all changes along with the reproducibility of the build process.

SLSA is maintained by the OpenSSF and provides lots of resources for adoption. It is a great framework for ensuring the integrity of software artifacts throughout the entire software supply chain.

Secret management

As described in *Chapter 8*, there are lots of secrets involved in continuous delivery itself, but also in the applications deployed by it. Therefore, you must ensure that they are managed in a secure way. This means that you should avoid storing secrets in plain text and in your source code repository. Instead, you should use a secret management solution such as HashiCorp Vault or AWS Secrets Manager. These solutions will help you to ensure that your secrets are stored securely and that you can easily rotate them. Furthermore, you should also ensure that you are not exposing any secrets in your delivery process, especially by logging secrets. This will help you to ensure that your secrets are not exposed and won't be leaked.

Immutable infrastructure

Immutable infrastructure is a very important concept when it comes to security. The idea behind immutable infrastructure is that you do not change your infrastructure after it has been created. Instead, you create a new version of your infrastructure containing the changes and replace the old version with the new version. This will help you to ensure that your infrastructure is always in a known state and that you do not introduce any security issues by changing your infrastructure. It will also make your infrastructure easy to reproduce when you need to debug an issue.

In this section, we discussed a few best practices for securing your delivery process. In the next section, we will discuss some best practices for the cloud provider's environment.

More best practices

Most cloud providers have their own best practices for DevOps and continuous delivery. These best practices are usually based on the experience of the cloud provider and are great resources to find out more about how to implement processes in the respective cloud. Here we present to you a few public best practices to help you craft your own journey toward continuous delivery with the cloud:

- AWS Well-Architected Framework/AWS DevOps Best Practices
- Azure DevOps Best Practices
- GCP DevOps Best Practices

Furthermore, the CDF has a best practices website (https://bestpractices.cd.foundation) with lots of best-practice resources for continuous delivery covering different topics such as security, testing, monitoring, and more. These best practices are usually based on the experience of the community and are great resources to find out more about how to implement processes in your organization.

Case studies

Case studies are illustrations of success stories, where a detailed examination is made of the problem, its solution, and the associated best practices that came from this. Case studies provide successful evidence-based implementations of continuous delivery in cloud-based ecosystems. Here we present to you a few public case studies to help inform your journey with continuous delivery on the cloud.

Case study 1 – Coca-Cola Argentina app development and product delivery

Coca-Cola Argentina delivered soda products to millions of South American customers. Argentina was one of the largest distribution regions with a number of production plants and bottling companies. These small businesses accounted for 75% of the overall sales. Keeping track of inventory, delivery channels, ordering, and lead times was a huge challenge. A digital transformation of the whole supply chain was ongoing and speed and agility were critical for the success of the overall initiative. The company decided to move the application development environment to a cloud provider.

A mobile app was developed that ran on the cloud with AWS CodeDeploy (which helps to automate, configure, orchestrate, and monitor software deployment – https://aws.amazon.com/codedeploy/) to automate Coca-Cola's software deployment.

Coca-Cola achieved greater efficiency with a microservice-based architecture for its mobile application and this enabled more cost-effective scaling along with the services provided by the cloud provider. This also helps to incrementally deliver features with speed and resilience. Additionally, the mobile application provided quick access to inventory for consumers, be it store owners or direct B2C consumers. Products ordered through the mobile application can reach the consumer in minutes instead of hours and days.

More information about this case study can be found at `https://aws.amazon.com/solutions/case-studies/coca-cola-case-study/`.

Case study 2 – automate deployments with CI/CD pipelines on Azure DevOps

In this scenario, a client already started their DevOps journey and already had a containerized application. The application was moved to a cloud provider and Azure resources and services were used to do this migration. An Azure DevOps pipeline was used to deploy the infrastructure and the application. A managed **Azure Kubernetes Service** (**AKS**) instance was used to migrate the application to the cloud provider with minimal configuration changes. Azure Application Gateway, a managed Azure MySQL database, a monitoring application, and an Azure Log Analytics workspace were the other services used to migrate the application stack.

The client's environment became more scalable and efficient because the AKS environment gave them the ability to scale up and down based on the demand. Furthermore, the build and deployment processes were streamlined and automated. All code changes to the web application were triggered through the build pipeline; the pipeline compiled the application for release and pulled the container image to build pre-deployed containers. The Azure release pipeline pushed the relevant Helm chart to the AKS cluster and subsequently started all defined components. This automated and managed deployment ensured consistency, reliability, and ease of deployment at all steps.

The original case study can be found at `https://www.rackspace.com/sites/default/files/2021-07/AKS-DevOps-CaseStudy1.pdf`.

Case study 3 – developing complete end-to-end CI/CD pipelines at Ericsson

The key challenges and improvement opportunities referred to in this case study are aligned with the author's vision, detailed in *Chapter 11*. In large organizations, creating alignment and synergies for tools, processes, and event technology choices is not only difficult but also inefficient.

A highly complex end-to-end software flow for multiple intertwined products is a new reality for many large incumbents, for particular tools used by Ericsson refer to the case study document (found here: `https://cd.foundation/wp-content/uploads/sites/78/2023/04/CDF_Ericsson_User_Story_May2023.pdf`). Continuous delivery itself for such a complex ecosystem is a huge challenge. Scalability, traceability, and observability are critical to ensure the deployment of reliable and maintainable software. Some of these aspects are aimed to be solved through innovative ways as described in the case study as follows (detailed in the public case study):

- Using a common standard events as binding CI/CD pipeline and ensuring interoperability

- CDEvents aims to create a common specification for CI/CD events

- Common event protocol natively supported by different open source tools such as Jenkins, Spinnaker, Tekton, and any other tool in the CDF landscape

Furthermore, the number of interaction points and interfaces between many tools in the continuous delivery ecosystem creates challenges for practitioners. The case study discusses a progressive posture for continuous delivery and related pipelines with event-driven capabilities integrated through open source tools. The benefits of using events for interoperability include avoiding vendor lock-in, minimizing the effort to employ new tools and capabilities seamlessly in the pipelines, and allowing events to be used to productize pipelines to reduce the cognitive load on practitioners.

These case studies are just a few examples of how companies are using continuous delivery in the cloud. There are many more case studies out there in which you can find more examples of this topic. Let's now summarize what we have learned in this chapter.

Summary

In this chapter, we highlighted some best practices that will help you to improve your delivery process. First, we discussed what best practices are and why they are important. Then we discussed some best practices for the whole application deployment process, the delivery process, and specifically for securing your delivery process. We also discussed some best practices from cloud providers and reviewed some case studies. We hope that these best practices will help you to improve your delivery process and to achieve your goals. At this point, we want to thank the community for their contributions of best practices to this chapter. In this chapter, we also highlighted a few ways to accelerate adoption as well. Finally, it is important to continuously educate yourself and others and get buy-in for such initiatives; best practices can be improved, and continuous improvement by searching for new ways forward should never stop.

Chapter 10 will cover future trends for continuous delivery in the cloud.

10

Future Trends of Continuous Delivery

We covered many aspects of modern technologies that are constantly evolving the posture for **continuous delivery (CD)** in previous chapters. Through advancements, there has been substantial progress made when it comes to the stability and throughput of software delivery. Referring to the results from the *State of DevOps* report over a decade, organizations have adopted different aspects of CD practices starting from cloud technologies, automation, and progressive deployment.

Organizations also measure progress using different types of metrics; most of them have adopted **DevOps Research and Assessment (DORA)** metrics in some way. However, with the evolution going mainstream, it is also becoming complex, and practitioners are constantly being challenged to learn about new products, tools, and new technology. Providing some context to this chapter, the evolution of CD is driven by the challenges and constantly changing ecosystem of producers and consumers of CD.

We will cover the following topics in this chapter:

- Key challenges of the current state of CD
- Functional advancements
- Architectural advancements
- Operational advancements
- Adoption challenges and key recommendations

Key challenges of the current state of CD

Before we try to simulate the evolution of CD through this chapter, let's try to establish the baseline or the purpose of this evolution. Of course, some aspects might be random, but most of the parts that are discussed in this chapter will be evolving due to the current challenges and constraints introduced because of architectural or functional limitations. There are also operational and process aspects that might be a bottleneck in the future. So, the following are some identified challenges:

- **Exponential growth in tools**: It is hard to believe that we only started a decade or so ago with most of the practices and concepts in the CD space. Today, hundreds of tools are available that can perform foundational tasks for CI/CD.

- **Co-evolution of practices, tools, and the cloud**: The co-evolution of processes and tools does interrupt stability with bursts of short evolution cycles, whether it be **Infrastructure as Code (IaC)**, progressive delivery, or open source contributions. New tools are created and evolved in a relatively short span, introducing a change curve for the community.

- **Complexity**: It is inevitable that complexity is increasing, with rapid adoption, and there are new risks, operational overhead, and cognitive load on practitioners and organizations. Where we started from, with CI as a simple solution to a specific problem, has now grown into an ecosystem of CD, with a number of aspects working together.

- **Cost**: With more and more services and applications moving to the cloud, financial practices and engineering practices are getting tightly integrated. Cost optimization, simplified chargeback, and better provisioning and utilization of resources are becoming integral parts of the CD ecosystem.

- **Skills shortage**: There are many aspects of learning, and many people are getting left behind due to a lack of time for self-development; this is introducing a major skills shortage. Educational institutes, certifications, and other bridging means for skills enhancement are also having a hard time catching up.

In the next section, we will introduce some aspects of functional, operational, and architectural evolution, and we will discuss some progressive ideas around these advancements.

Functional advancements

Many aspects discussed in the previous section indicate an alignment with the functional advancements of CD to simplify the posture of tools and processes. Let's look at some of these:

- **Interoperability within the cloud ecosystem**: With the co-evolution of tools and the cloud itself, there is a need for seamless migration of workloads across different cloud providers and different landing zones. Organizations are looking at a seamless transition without requiring huge efforts to re-configure, test and re-synchronize the CD workflow when these migrations take place. The functional *super-cloud*—sometimes referred to as the *cross-cloud,* providing interoperability and, especially, CD capability—is one progressive area of functional evolution. It is expected that applications and tools will work on orchestrating higher levels of interoperability, and standards will play an important role.

- **Event-driven CD**: Using an event-driven approach introduces a high level of reusability, flexibility, and full-stack interoperability for the complete software life cycle. The primary idea is around *declarative events* that act like a common language, where the producer of the event sends information about the event without knowing how the consumer will use this further. So, the CI/CD workflow will be made of tools that can produce and consume these declarative events in a seamless manner without static and imperative definitions. Examples of common sets of abstractions, as defined in CDEvents, an incubating project under the **Continuous Delivery Foundation** (**CDF**), are the following:

  - **Core events**: Events such as pipeline runs or task runs

  - **Source control events**: Events for source code assets

  - **Continuous deployment pipeline events**: Events related to the environment, covering where artifacts are produced by CI pipeline runs

  - **CI pipeline events**: Events related to building, testing, packaging, and release artifacts

- **Future of progressive delivery**: Combining progressive delivery with **machine learning** (**ML**) capability and reducing the challenges of adoption of progressive delivery. Cutting the long tail of technical debt created through a number of feature flags and associated code versions or blots. Keeping track of the performance of the feature flags.

- **Security**: In response to the rise of open source technology integration and cloud hosting of pipelines for software delivery, there is also a specific investment into improving the security posture. We covered at length security for CD in the previous chapters. In this chapter, we highlight the evolution of security considering future perspectives. If we relate security to the *super-cloud* or the *cross-cloud*, there is a significant architectural shift. On the other hand, if you assess the security from an open source integration perspective for CD, a lot more will happen in the **software bill of materials** (**SBOM**) and the security of the overall software supply chain. Here are some things that are happening currently:

  - **Open Cybersecurity Schema Framework** (**OCSF**): This is an initiative by AWS and Splunk. Among other big companies, IBM, Cloudflare, and Salesforce are also backing it up. This framework is focused on data in the first place. The OCSF community is actively working toward streamlining security operations.

  - **Cloud-native application security**: From container scanning and security control tools to IaC misconfigurations, there is a lot of action in the security space when it comes to application security. By 2025, over 95% of new digital workloads will be deployed on cloud-native platforms.

  - **A silver bullet for the SBOM of the future**: As the SBOM continues to evolve, so will the framework for data interchange and the need for a standard format. As discussed in the previous chapters, the three prominent formats are **software package data exchange** (**SPDX**), CycloneDX, and **software identification tags** (**SWID**) tags, so vendors are looking to integrate at least two of them at this stage. In the coming years, more and more capabilities and decision-making features will be integrated with the support of **artificial intelligence** (**AI**).

CD will undergo continuous improvement, with more standard and automated components integrated into the entire life cycle. In the next section, we will address the architectural evolution and industry perspective on the evolving capabilities of tools and applications accordingly.

Architectural advancements

Some more disruptive architectural advancements are likely to transform how we develop and deploy applications and even go beyond cloud-native. Before we dig deeper into the architecture, let's discuss key trends for next-generation applications from a micro perspective:

- **Portability**: A variety of programming languages, platforms, and frameworks
- **Observability**: Decentralized and distributed management
- **Resource-optimized and resilient**: Optimized posture of infrastructure
- **Real-time and dynamic**: Support rapid scale-up and scale-down and address real-time requirements

There are more macro trends or big-picture outlooks that are also influencing the architecture when it comes to cloud-native applications. Let's also discuss the macro changes:

- **Cloud-native to edge-native applications**: Edge-native applications have cloud-native principles while taking care of resource usage, latency, bandwidth management, security, portability, and reusability of applications. A rise in edge-native applications would likely influence trends and investments for architectural advancement.

- **Experimentation with WebAssembly (Wasm) applications**: Wasm is a lightweight, fast, and secure *container* for functions. It is a binary instruction format and a virtual machine that allows developers to develop applications in their choice of language and supports the new approach of edge-native. The rise of Wasm looks promising; however, it is still in its early stages. Projects are experimenting with running Wasm on Kubernetes, which acts as a default *control plane* in many cases. Microsoft **Azure Kubernetes Service** (**AKS**) has also announced the integration of Wasm.

- **Declarative APIs for distributed applications**: APIs are simplifying the deployment of distributed applications. API management and security is another area that will influence how the landscape of tools and frameworks will transform. APIs will continue to abstract complexity and the common challenges when developing applications in a secure and resilient way. Another aspect that will gain interest is API security itself and the tools and practices around it.

- **Pipeline-less architectures**: The preceding trends also indicate that the software development itself will transform into a more event-driven workflow where pipelines are an integral part but abstracted from the developers. Events will have to be standardized, and tools will be able to integrate in a seamless way.

In the next section, we will discuss operational advancements, keeping in view the explosion of tools and applications in the CD space. The operational overhead for DevOps engineers and site reliability engineers is increasing, and it is important to address the operational advancements so that we keep on decreasing the cognitive load on our engineering staff.

Operational advancements

Many technological developments for CD are intertwined. Keeping an eye on operational complexity, resilience, and reliability also becomes critical. In this section, we will talk about some of the key developments that are steered backward from production to development:

- **Simplification of the operational posture**: To embrace incremental technology advancements for CD, it is important to double down on standardization and simplification, with the integration of automated and repeatable processes to address the reliability needs of cloud-native applications in production. Due diligence is required to keep the operational posture clean, simple, and optimized.

- **Next-generation service management**: While we advance through modern CD capabilities into our applications, it constantly shifts our perspective to portable and reusable components. To ensure a robust and resilient posture of production-ready applications, they must be portable, interoperable, and cloud-agnostic. It is no longer about using one single tool for **Information Technology Service Management** (**ITSM**), change management, or incident management but more about how we can get one unified view of critical events through practices such as observability.

- **Showback and chargeback**: Monitoring cost and resource utilization at the application or even functional level instead of at the **business unit** (**BU**) or organizational level. The transition would need the integration of new tools and expertise to manage cloud-native workloads.

- **AI centricity**: This capability is in the formative phase; there is a lot of excitement about ChatGPT, OpenAI, AIOps, and ML in general. Whether predictive analysis, anomaly detection, or conversational bots, AI has the potential to revolutionize operations and related jobs. Practically implementing these solutions would require the training and tuning of models on a constant basis and will give rise to new tools, applications, and job roles such as data scientists and ML engineers becoming part of the organization's operations. As this space continues to evolve, we watch out for key developments in this area.

- **A system of service-level objectives (SLOs) and service-level agreements (SLAs) integrated with** CD: There is more work needed to ensure QoS and a good customer experience when it comes to cloud-native applications. Integrating quality sprints into CD has a trade-off; often, organizations struggle with cost and time implications and do at times oversee the value of such sprints in a short time frame. To overcome the problem, Google came up with a system of SLOs, SLAs, and error budgets. This is also often adopted by various organizations and is part of **site reliability engineering (SRE)** practitioners' scope; however, it is extremely difficult to maintain the cadence of error budgets with SLOs and SLAs in complex and large-scale organizations. These principles and capabilities tend to fall off over time. In the future, we might see the simulation of SLAs and SLOs by introducing an ML capability. Designing such simulations can be a daunting task; we might see a rise in platforms that can be purpose-built for SLOs/SLAs and error-budget cadence and simulation. Today, there are discrete efforts to design such simulators through OpenAI Gym or SimPy.

With that, we conclude this section on operational advancements. The preceding discrete topics can be converged, and we will see incremental steps through projects and initiatives within the organizations as a first step.

Adoption challenges and key recommendations

We are advancing into the next wave of cloud-native technologies, and CD plays a pivotal role in enabling this. We discussed earlier some of the key advancements for the CD stack; however, there is an uphill road to adopt and scale future capabilities. We see organizations are becoming vigilant and continuing to watch out for *cloud exit plans, DevOps is dead, what next?*, and so on, partially due to adoption challenges but more because of lack of guidance and support. Let's discuss how to avoid such pitfalls for CD and cloud-native in general:

- **Cost tracking, management, and optimization**: It is important to be in control of the **return on investment (ROI)** of onboarding new technology, tools, and processes. The CD posture can dictate expenditure if this is not managed and optimized consciously. A third-party audit can be one way to keep checks and balances for the future.

- **Best practices and general guidelines**: We elaborated on the best practices for CD in the previous chapter. Best practices come out of a community of practitioners—people implementing CI/CD at scale, experimenting with ideas, and collaborating to improve them. These best practices have something in common: they are meant to save you from making big mistakes, guide you through the implementation of standardized components, and ensure that you have support when you need it. There are many communities constantly working to establish best practices, whether it be Kubernetes, observability, or CD. No one should have to go on reversing the path of advancement due to a lack of community support.

- **Tired of making hiring mistakes**: In general, it is hard to hire people for CD as it is a stack of practices, tools, and technology instead of specialized skills. In the quest of hiring the right people, an organization may lose track of the in-house talent pool and upskilling programs. Moreover, it also requires organizational change; often, more of it is making it safe to experiment and fail fast, not to leave behind the most important skill set of being innovative and open to collaboration to solve critical business problems rather than being a technology expert.

- **Unprepared for open source**: Learn more about open source initiatives and how they fit into your organization. Most recent developments are credited to open source initiatives, especially in the CD space. Going in *unprepared for open source* would be a big pitfall in the journey as it can lead to many constraints. Kubernetes and OpenTelemetry, modern CI/CD tools such as Argo CD, Spinnaker, and so on are gaining traction in the CD space. Ensure you embed processes and people to steer open source initiatives within your organization.

CD has been around for a while; however, it is still an evolving space. There is a lot of action, innovation, and, to a certain extent, complexity in predicting future trends. Keeping up pace with everything can be overwhelming; watch out for burnout and cognitive load on the people implementing CD. To conclude this section, we looked at some common pitfalls and adoption challenges for our readers. This will help practitioners to stay in sync with the evolving needs of the CD stack and shifting technology landscape.

Summary

In this chapter, we briefly touched base on functional, architectural, and operational trends, which can help us respond well to the evolution of CD. The heterogenous growth of tools and applications can be overwhelming, and it is bound to consolidate the overall CDposture to a more standardized one. Tough decisions will be required to coordinate the evolution and maintain the reliability and predictability of the dynamic ecosystem of CD.

In the next chapter, we will address the important topic of community contributors. We will discuss the various basic aspects of open source projects, various organizations that are fostering open source projects, and how you can contribute to the open source ecosystem.

11

Contributing to the Open Source Ecosystem

This chapter focuses on new open source projects for potential contributors. We will introduce these projects for further reading.

The chapter will cover the following topics:

- Overview of the **continuous delivery (CD)** open source ecosystem
- Open source licensing
- The governance model and maturity levels
- Sandbox, incubating, and graduating projects
- How to be a contributor to the open source ecosystem of CD
- Long-term sustainability
- Funding and investment in CD in cloud projects
- Key considerations for funding open source projects
- Vulnerability management

Overview of the CD open source ecosystem

In modern software projects today, there is a substantial portion of open source code. Organizations have realized the potential of open source. According to reports, proprietary software is no longer the first and only option enterprises are leaning toward. Primarily focusing on the cloud-native era, open source tools and technology are fueling innovation, whether Jenkins, Chef, Puppet, Ansible, Terraform, OpenShift, Cloud Foundry, Docker, Kubernetes, or OpenTelemetry, there is a common thread that leads to open source. All these innovative projects are steered through the collaborative effort of communities and individuals and supported by big and small organizations. Large enterprises are investing in the open source ecosystem and there is a shift in ownership as we move along. In 2018, Microsoft bought GitHub; similar other acquisitions are seen in different segments by industry incumbents. We also see that open source projects come of age and reach the maturity level for adoption at scale. Kubernetes is one of the great examples; this project has been instrumental in disrupting the infrastructure space. Some examples of leading open source projects at the time of writing are documented in the CDF Landscape (`https://landscape.cd.foundation`) and CNCF landscape (`https://landscape.cncf.io`). These two links refer to the full landscape of open source projects active in the CD ecosystem from these foundations. As we progress through the chapter, we will also introduce other organizations, foundations, and non-profit initiatives.

Some of the popular open source projects in the CD space are outlined in this table:

CI and Pipeline Orchestrator	Infrastructure Deployment
Argo (`github.com/argoproj/argo-workflows`)	Kubernetes
JenkinsX (`github.com/jenkins-x/jx`)	Pulumi (`http://github.com/pulumi/pulumi`)
Keptn (`github.com/keptn/keptn`)	Terraform (`http://github.com/hashicorp/terraform`)
Tekton (`github.com/tektoncd/pipeline`)	Ketch (`https://github.com/theketchio/ketch`)
	Helm (`github.com/helm/helm`)
	Shipwright (`github.com/shipwright-io/community`)

Table 11.1 – Open source projects in the CD space

Open source software is often developed in a decentralized way. The *2023 State of Open Source Report* provides insights into the key trends, adoption challenges, and overall state of open source software in general in organizations. As per the report, the top reasons for adopting open source are as follows:

- The use of open source software gives you access to innovations and the latest technologies.
- The functionality available to improve development velocity is a close second, strengthening the argument that nearly all software is built on top of, or with, open source. There are millions of open source libraries improving developer productivity.
- Rounding out the third reason is stable technology with community long-term support, which makes sense — organizations tend to favor proven technologies backed by active support communities.

When dealing with open source, we see that there are substantial benefits in adopting open source technologies as a key building block for the CD ecosystem.

However, there are some challenges as well. Whether it's ensuring adherence to security policies and regulations, addressing skill or experience gaps, staying proficient in the latest technologies, or keeping up with software updates and retiring end-of-life versions, there are many challenges that organizations face in maintaining their IT infrastructure.

The report also highlights that **software bills of materials (SBOMs)** are becoming more popular, with retail, banking, insurance, and financial services leading in this category and leading implementation projects for generating SBOMs.

Gartner, in its research, has highlighted the main challenges with **open source software** (**OSS**), primarily open source licensing, governance model, sustainability, and vulnerability management. Let us look at them in detail:

- **Open source licensing**: As OSS projects grow in popularity, so do the number and types of licenses. A license serves as a binding contract regarding how software can be used, with **terms and conditions** (**T&Cs**). Broadly categorizing the licenses, either they are copy-left or permissive. An OSS project using a copy-left licensed software component will in general only be allowed to use it; if it can be used as open code, then it can be used by others. However, permissive licenses allow projects to use the open source component for proprietary work as well, under the T&Cs outlined in the license. However, these broad categories are only the tip of the iceberg, as there are hundreds of license types originating from these, and the problem of finding a sweet spot for a business model still remains a challenging task for any of these initiatives.

- **The governance model and operative structure**: Starting from structure and governance of the OSS projects to adopting best practices, there are a lot of aspects to developing an effective operational model for OSS projects. Defining roles—for example, maintainers, contributors, and reviewers—and developing a governance structure that outlines how open source projects take critical decisions, create roadmaps, and even formalize leadership are some of the aspects. There is more work needed in order to ensure strong operational governance to keep the community around the project motivated.

- **Long-term sustainability**: Open source projects have contributed to filling an innovation gap, especially when it comes to the CD ecosystem; however, the sustainability of these projects has been a challenge. Right from funding to community engagement, there are a number of obstacles. We will cover some aspects of overcoming the hurdles in this chapter.

- **Vulnerability management**: As OSS projects are driven by the community, it is important to develop and maintain a set of best practices regarding how the project manages vulnerabilities, primarily identifying, addressing, documenting, and automating the workflow to provide an understanding of the dependencies and impact in general. A number of resources and open source tools are available to support these projects and the communities behind them, but it still is hard to manage vulnerabilities for a number of reasons, including a decentralized approach to developing the projects and a lack of software supply chain visibility and consistency in addressing vulnerabilities.

Open source licensing

There are many myths about open source, which we will also uncover in this chapter, starting with *why open source licensing*? Let's start with the basics; one of the biggest misconceptions is that open source is often considered just free access to a code base. Open source licenses provide the terms of use, distribution, and modification, with obligations defined around open source software, including the code base. It is considered a contract between the owner, contributor, and distributor in one way. Open source software is considered a disruption from the way software was developed in the past as **intellectual property (IP)**. Primarily, open source licenses are categorized into two broad categories, **copy-left** and **permissive**.

Copy-left ties back to the free software movement where you are allowed to use, modify, and distribute the software with reciprocity, which means any software that intends to use copy-left will be distributed under the same type of license. A permissive license provides strict criteria as to how the software can be used, modified, and distributed. Licensing itself is becoming complicated as it evolves from the broad categories, as mentioned earlier. The most popular license types are as follows:

License	Type	Reference
Apache license	Permissive	`https://www.apache.org/licenses/LICENSE-2.0.txt`
BSD license	Permissive	`https://opensource.org/license/bsd-3-clause/`
MIT license	Permissive	`https://tlo.mit.edu/learn-about-intellectual-property/software-and-open-source-licensing/open-source-licensing`
Eclipse public license	Weak copy-left	`https://www.eclipse.org/legal/epl-2.0/`
Mozilla Public License	Weak copy-left	`https://www.mozilla.org/en-US/MPL/`
GNU's General Public License	Copy-left	`https://www.gnu.org/licenses/gpl-3.0.html`

Table 11.2 – Types of open source licenses

Popular projects in the CD space such as Kubernetes, ArgoCD, Keptn, Istio, and many more use the Apache 2.0 license. The popularity of permissive licenses, particularly Apache, has grown in recent years.

References and further reading material can be found on the **Open Source Initiative (OSI)** website (`https://opensource.org/osd/`).

In the next sections of this chapter, we will look deeper into the governance model and related challenges with the entire ecosystem.

The governance model and maturity levels

Open source is embedded into nearly every software produced today through direct or indirect means.

We now take a deeper plunge into CD and cloud technologies, which have enabled several modern software practices. Many popular areas such as containers, service meshes, microservices, immutable infrastructure, and declarative APIs are getting contributions from open source communities now more than ever. There are some focused organizations that have helped the community to broaden the impact by providing stewardship to the OSS projects.

Stewardship to the OSS projects

We will highlight a few such organization models and associated organizations with a focus on their efforts to grow the CD in the cloud open source ecosystem and provide guidance on the governance of these software projects in general:

- **Individual companies or industry consortiums**: Big companies often choose to distribute software under the terms of an open source license. The contributions of companies such as Google, RedHat, and Microsoft are particularly notable.

- **Contributor License Agreements (CLAs)** are one of the ways to grant rights over all contributions; these are often adopted by specific organizations. Under this model, the governing organization may not accept contributions from anyone outside it or require a CLA to accept a contribution.

- **Founder-leader**: The founder-leader governance model is most common among new projects or those with a small number of contributors. Often, these projects require support from different communities. Also, at some point in time, based on the size of the project, a council or board is established to govern the project.

- **Non-profit foundation**: This governance model is popular and is adopted by leading open source projects as, irrespective of size and funding, these projects can join under one umbrella and take advantage of set processes, resources, and a community.

Some organizations that are steering the open source movement for open source CD projects in particular are as follows:

- The **Cloud Native Computing Foundation (CNCF)**: As per the CNCF charter, it intends to serve a role in the open source community to drive the adoption of this paradigm by fostering and sustaining an ecosystem of open source, vendor-neutral projects and democratize state-of-the-art patterns to make these innovations accessible to everyone. Key trends are as outlined by the CNCF survey report. The CNCF is also starting to see organizations move up the stack—adopting less mature but innovative projects to tackle challenges, including monitoring and communications.

- The **Continuous Delivery Foundation (CDF)**: It hosts a number of projects in the CD space. However, the organization covers a broad range of initiatives in the CI/CD space. Right from fostering contributions to open source projects, it provides sustainable support to upcoming vendor-neutral open source initiatives. It also has a community of contributors, through which it is constantly evolving the industry specifications and reference architecture and also ensuring guidance through collaborative workshops and case studies. The initial open source projects hosted by the CDF were Jenkins, Jenkins X, Spinnaker, Tekton, Ortelius, CEevents, `screwdriver.cd`, Shipwright, and Pyrsia. The key trends as outlined in the *State of Continuous Delivery Report (2022)*, published by the CD foundation and created by SlashData, which is licensed under the Creative Commons Attribution-NoDerivatives Licence 4.0 (International), are as follows:

  - A strong correlation was depicted in the survey between speed and stability

  - A rise in the adoption of CD practices was seen particularly for medium-sized businesses

  - It was also outlined that there is more work needed when it comes to the automation of practices

  You can check out the recent report of 2023 produced under the same license, published by the CDF and created by SlashData:

  - It reflects that the adoption of DevOps practices is still on the rise

  - The velocity of code development might be suffering due to complexity, which reflects the need to simplify the posture of CI/CD tools in general

- **Apache Foundation**: Apache Foundation was established in 1999; it hosts around 300 open source projects with contributions from more than 8,000 open source community members. Some of the successful projects under its umbrella are Spark, Cassandra, Kafka, and Hadoop, to name a few. The **Apache Project Maturity Model** is a framework for evaluating the overall maturity of an Apache project community and the quality of the code base. The **Apache Software Foundation (ASF)** uses this framework for various open source projects under its umbrella. It also has community mentors and best practices outlining the adoption of the framework.

- **The Cloud Foundry Foundation**: The Cloud Foundry Foundation supports platform-as-a-service offerings for the open source ecosystem, specifically various open source projects, open data projects, and open standard projects. The ecosystem has certified distributors, certified system integrators, authorized training providers, and a number of infrastructure providers.

- **Cloud**: There are many attempts to provide an open cloud to users. Leading providers in this space include Open Cloud by Google, which has partnered with a number of open source providers with the ambition to provide a managed open cloud to end users. Apache Cloudstack is another attempt to provide a stack of features specifically for compute orchestration. An open source cloud initiative can provide an alternative to commercially owned cloud environments.

In this section, we highlighted some of the important players in the open source ecosystem when it comes to CD in the cloud. There are a number of foundations working on OSS projects, and many of these foundations have well-defined processes, especially for hosted projects. Google, Netflix, Twitter, and Facebook are leading technology companies that have shared their in-house tools and plugins to enhance the projects with the open source community.

On top of this, open source collaboration has been enhanced through platforms such as GitHub, which provides a robust way to share code and collaborate. So now, in the next section, we will describe the governance and the maturity model of the projects.

Sandbox, incubating, and graduating projects

Let's take an example from the CNCF, which is one of the leading organizations when it comes to supporting cloud-native open source projects with community members, research, events, and funding initiatives. Every CNCF project has an associated maturity level:

Project stages

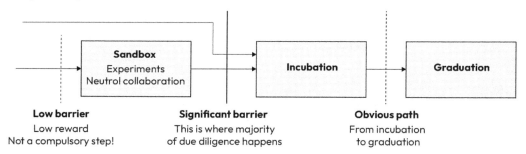

Figure 11.1 – Project stages of an open source project (Source: https://github.com/cncf/toc/blob/main/process/project_proposals.md)

Let us look at these stages in detail:

- **Sandbox projects**: The sandbox projects can be considered to be in an experimental stage. Some of the popular projects that were accepted as sandboxes in 2022 and 2023 include Kubescape, Opencost, Ko, and so on. These projects benefit from the CNCF community and a collaborative approach to further develop these projects. You can find the list of sandbox projects here: `https://www.cncf.io/sandbox-projects/`.

The following workflow diagram depicts how a CNCF sandbox application flow works:

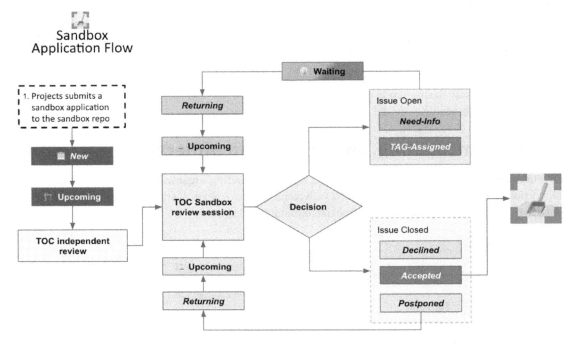

Figure 11.2 – Sandbox application flow (Source: `https://github.com/cncf/toc/blob/main/process/sandbox-application-process-2022.png`)

Projects apply for a sandbox through the Sandbox Repo: `https://github.com/cncf/sandbox/issues/new`. More information on this process is found on the main Sandbox repo page.

A further reading reference can be found here: `https://github.com/cncf/toc/blob/main/process/sandbox.md`.

- **Incubating projects**: Projects at this stage must meet the sandbox stage requirements plus many other criteria, which are outlined in the process document – some of the main considerations being the number of committers, the number of commits and contributions, versioning, the documented security process, and public reference for implementation. The metrics and the criteria can vary from project to project depending on scope, size, and type. The **technical oversight committee** (**TOC**) has the final judgment on the outlook and prospects.

The following diagram depicts the workflow for an incubating project workflow:

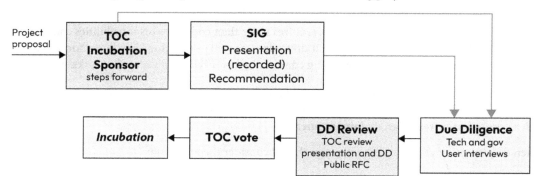

Figure 11.3 – Incubating project application flow (Source : https://github.
com/cncf/toc/blob/main/process/incubation-process.png)

- **Graduation stage**: To graduate from a sandbox or incubating status, or for a new project to join as a graduated project, a project must meet the incubating stage criteria plus some more criteria added, mainly reflecting the adoption of the project by organizations. There are badges associated with the maturity and it is included as part of the graduation checklist that the project has earned certain badges. The project governance and committee process is also essential; however, it is necessary to have a super-majority vote from the TOC to move to the graduation stage.

Further reading can be carried out at `https://github.com/cncf/toc/blob/main/process/project_proposals.md`.

How to be a contributor to the open source ecosystem of CD

The open source ecosystem provides an immense opportunity for everyone to contribute. The community plays an important role in providing mentorship to newcomers. Open source projects that have diverse contributors from the user base have solid foundations.

Let's discuss some of the opportunities and how they help the community and its projects:

- **Organic innovators**: The developers contributing to OSS are responsible for adding new ideas and features.

- **Advocates**: An open source ecosystem consists of a community of developers, users, and researchers who also act as developer advocates for the OSS.

- **Developers**: If you are a developer, there are many ways to contribute, basically starting with writing the code, and if you find bugs, then you can make a pull request to fix it.

- **Maintainers**: Maintaining a project requires more than code. The responsibilities can apply to the documentation and keeping it up to date, as well as defining workflows and rules to contribute. Mentorship and leveraging community contributors are some key tasks.

As a part of governance, all projects have roles, even if they are not explicitly documented. Documenting them will help you recruit people for those roles. A minimal set of project roles will often be something such as **contributor**, **reviewer**, and **maintainer**.

When documenting your project's roles, you should consider the following:

- Why do you need contributors and do you have defined roles?

- Do you have guidelines for becoming a contributor? Sometimes, it is also popular to create a program to track the contributions and issue badges.

- How can these contributors interact with other community members and associated authority?

- Does the community meet at a regular cadence so that contributors can participate?

- What project resources are the province or responsibility of people who perform certain roles?

In the next section, we will discuss aspects to consider for the long-term sustainability of projects.

Long-term sustainability

As highlighted earlier, there can be several factors that influence sustainability; we will start with one of the most important considerations, **economic viability**.

A **business model** is a systematic way of creating long-term value by monetizing a product and its features. On the other hand, an OSS licensing model and the collaborative way of developing applications have given rise to a whole industry on top of which business models have been built, which we call **open source business models**. One question you might have is how these companies monetize OSS. The rise of monetizing open source has grown leaps and bounds with DevOps, and different types of business models have been adopted. One way to visualize the worth of commercial open source companies is through the COSS Index. It is a public Google sheet and is accessible here: https://docs.google.com/spreadsheets/d/17nKMpi_Dh5slCqzLSFBoWMxNvWiwt2R-t4e_17LPLhU/edit#gid=0. Let's have a look at a few examples of business models adopted by different open source projects to understand this better.

Let us look into various models through which different companies have monetized open source:

- **Support and services model**: OSS can also be commercialized in terms of support and services. Another possibility is offering OSS in source-code format only, supporting paying customers with executable binaries, compiling, and packaging. An example of such a business model is

Red Hat. Through the software subscription option for Enterprise Linux, Red Hat has been able to monetize the extensive validation and testing of Linux releases and also has offered an enterprise version hosted on several cloud providers. Canonical offers software for free and charges for professional services. Similar references can be found for Spinnaker, a widely adopted open source tool for continuous delivery. Companies such as Armory are extending open source to enterprises by providing systematic approaches to scale the adoption through many different types of subscriptions.

- **An open-core model**: Coined by Andrew Lampitt in 2008, this commonly refers to the core component or feature-limited software being free and open source and additional or add-on features being commercial. GitLab is such an application that uses open core model, with a hosted service offering that is free and open. However, the software for GitLab is closed. Kafka, Cassandra, Eucalyptus, Puppet, and Neo4j are some more examples.

- **Partnerships**: One popular example is GitHub, a software development and version control application. This application is popular for enabling collaborative source code development within the developer community. Microsoft acquired GitHub to integrate it into its enterprise offering for approximately $7.5 billion in 2018. With the marketplace offering, GitHub expanded its app development capabilities and was able to monetize them further.

Funding and investment in CD in cloud projects

Open source projects often start with a very specific problem, which means a narrow scope. Finding funds for those projects can be a challenging task. Sustaining open source projects is also a difficult task. Funding is an essential part of the success of an open source project. We present to you some guidance based on successful open source initiatives:

- **Project-to-product**: Projects such as GitLab have experimented with the project-to-product approach. This approach can help in funding the initiative by creating a SaaS offering that helps bring subscription-based funding to projects. Product managers work together with developers to bring incremental features in small chunks for rapid innovation and better reliability.

- **Sponsorships and FOSS funds**: According to recent articles and blogs on funding topics, it seems investing in open source is the new normal. Big companies have established FOSS funds, such as Salesforce, which was one of the first companies to launch a FOSS fund (built at Salesforce). Microsoft launched its own FOSS fund to support open source projects it uses and, more recently, those investing in diversity, equity, and inclusion. One such program run annually by Google is **Google Summer of Code** (**GSoC**); this program provides stipends to contributors who successfully complete FOSS projects in the summer.

- Here are some further references to the funds:

  - **Salesforce FOSS fund**: `https://engineering.salesforce.com/open-source/`
  - **Microsoft FOSS fund**: `https://github.com/microsoft/foss-fund`

- **Crowdfunding**: When a project or venture raises money from many people, typically via the internet or through similar social channels, it is termed as crowdfunding. LFX (the Linux Foundation crowdfunding tool) supports a number of open source projects in the CD and cloud-native space. Through this initiative, new open source projects are supported as often open source projects struggle with financing, legal support, training, project management, and technology infrastructure. LFX tries to solve some of these issues by providing one platform.

In the next section, we will discuss how open source projects are funded and the key considerations.

Key considerations in funding open source projects

As much as we want to consider that all open source projects are the same, investors find pitfalls with that idea. The following considerations and lessons learned will help reduce confusion about which projects are preferable to fund for the organizations who have set funds aside or investors who are likely to invest in open source projects:

- **Community**: The open source project is more viable if it has individual and institutional contributors.

- **Project usage**: The broad criterion is if the project has actual users.

- **Project license**: There are hundreds of open source and free software licenses, each with its own nuances. Check for the most widely accepted and curated list. Two examples are OSI-approved licenses, `https://opensource.org/licenses`, and Debian Free licenses, a curated list of which is available here: `https://www.debian.org/legal/licenses/`.

- **Project funding**: The projects should maintain information in a transparent way about the funding received and the sources. Gaining the support of non-profit foundations can be one way to go, to keep things neutral and options remain open. One other thing to consider is whether the project already has strong corporate funding and does not require any additional support.

- **Project ownership**: There are legal constraints at the company level to fund projects owned by an employee. Funding a project means supporting the leadership, governance, and structure of the project, so careful diligence is needed once you put that in place for your open source initiative.

Examples of well-funded open source projects

Let's check out some open source projects that are well funded and widely adopted:

- **Kubernetes**: Kubernetes was owned by Google and was managed through the community. It was released as open source in 2014. Google Cloud is referred to as the birthplace of Kubernetes by various experts. By 2018, it gained enough traction and was donated to the **CNCF** and Google provided approximately $9 million as per industry reports to support, run, and develop the Kubernetes project further; this funding was divided into a timeframe of three years.

- **OpenTelemetry**: OpenTelemetry is a combination of OpenTracing and OpenCensus' large, healthy open source projects. OpenTelemetry is used by hundreds of companies, including large enterprises such as Microsoft, Facebook, Netflix, Uber, and Twitter. Google is actively contributing to it. Google's contribution to OpenTelemetry includes the following:

- **Prometheus**: Prometheus is the primary library for OpenTelemetry. It provides the infrastructure needed to build OpenTelemetry-compatible systems.

- **Time Series Database (TSDB)**: Stores metric values and their corresponding label names.

- **RESTful API**: Allows querying TSDB using HTTP requests.

Microsoft has worked on OpenTelemetry since 2014. Microsoft's contribution to OpenTelemetry can be divided into four categories:

- **Prometheus**: The main library used by the project

- **Grafana**: A visualization dashboard that displays metric graphs and charts

- **Prometheus Operator**: Deploys Prometheus

- **JsonWireProtocol**: The protocol used by Prometheus for communication

Facebook has invested heavily in OpenTelemetry in recent years. There are two types of Facebook donations to OpenTelemetry. The first category consists of the core libraries. These libraries provide the infrastructure needed to build an OpenTelemetry-compatible system. The second category consists of the projects that use these libraries. There are many other big companies, such as Splunk and Datadog, and individual contributors that make these projects powerful.

Vulnerability management

This topic was discussed in detail in *Chapter 7*. Vulnerability management is one of the essential parts of open source projects. There are a number of best practices available that contributors and projects can consider to strengthen the security posture of an open source project. It is important the projects consider the adoption of these best practices and continuously identify, prioritize, and address vulnerabilities.

There are also a number of open source projects for vulnerability management, typically taking an example from the CD ecosystem:

- **Open Policy Agent** is a graduated CNCF project, which provides policy-based control for cloud-native environments

- **Sigstore** enables developers to sign software artifacts and much more

- **Ortelius** is an ppen source supply chain catalog that unleashes DevOps and security intelligence siloed across containers and pipelines

There are focused efforts to enhance open source security. The **Open Source Security Foundation (Open SSF)** is one such organization fostering collaboration upstream and allowing open source communities to advance open source security for all.

Summary

If you would like to start an open source project, this chapter provided you with the key considerations to get you started. As mentioned earlier in the chapter, the project must solve a specific problem, and there are various communities and channels in which you can discuss and collaborate with people to assess the problem statement itself. Once you have built some ideas on how to solve the specific problem, it's time to think about the next steps. Primarily, every open source project has a project page with a README file that explains how to use your project and also the purpose of the project. Another key aspect is the license, which allows others to use, or modify the source code. As a best practice, you have to add this file to the repo. Examples of popular licenses for open source projects are the likes of MIT and Apache 2.0 GPLv3. Lastly, consider the contribution guidelines or similar artifacts that can help contributors. Software versioning when working on software projects, major, minor, and patch releases are commonly used syntax for versioning. One final important thing to consider is the code of conduct; this is basically an agreement between all the contributors that they will behave in a respectful manner. Open source projects are created by people who wish to collaborate in an open space. By becoming the creator of open source projects, you will be able not only to create something new but also welcome others to contribute to the space.

In the next chapter, we will introduce a case study that aims to test the knowledge gained throughout the various chapters of this book. The case study will help you assess your knowledge and also implement the teachings. The case study assignment is aimed at critical thinking and applying the concepts introduced in this book. Readers are encouraged to explore possible solutions; it is advised to take time to read through the case study and connect the dots with the various concepts in this book with practical implementation.

References and further reading material

- *Open Source Projects - Beyond Code: A blueprint for scalable and sustainable open source projects* by John Mertic
- *Open Source Law, Policy and Practice* by Amanda C. Brock (editor)
- The collection of blogs, articles, and eBooks at `https://opensource.com/`
- *Open Source Governance Board* by Garima Bajpai: `https://ortelius.io/blog/2021/09/08/open-source-governance-board/`
- *Bootstrapping an Open-Source Community through Engagement* by Garima Bajpai: `https://ortelius.io/blog/2021/04/02/bootstrapping-an-open-source-community-through-engagement/`

12
Practical Assignments

This chapter focuses on practical assignments to test the knowledge of readers.

It covers the following topics:

- Overview
- Case study
- Task 1: Business alignment
- Task 2: Operational alignment
- Task 3: Technical architecture and strategy
- Task 4: Getting into the cloud
- Task 5: Advanced technologies
- Task 6: Security and compliance
- Task 7: Finding the tooling and necessary services to start to implement

Overview

As continuous delivery in the cloud has essentially disrupted our approach to designing, developing, testing, and deploying software applications, we would like to take a new approach to test the knowledge gathered through reading this book. We will be using a case-study-based approach where readers are expected to read and respond to the real-life problem by applying the knowledge gained from this book. One might ask: why not use the traditional question-based approach? The answer: to associate the learnings with near real-life scenarios that help bring forward critical thinking and creativity, and contribute to the professional growth of readers in a positive way.

As we know that there is no *one-size-fits-all* solution to such (case-study-based) problems, using such methodology for testing knowledge provides opportunities to brainstorm the case studies in groups and even build scenarios of different approaches and link the pros and cons.

The goal for readers is to build upon their knowledge and think practically about real problems and associated solutions based on their learnings from this book. Readers are encouraged to develop potential solutions to the open-ended case study. The case study can be discussed with like-minded readers in groups and communities to brainstorm potential solutions and share ideas. Now, let's look at an open-ended case study.

Case study

Before detailing the case study, let's try to summarize its elements for better understanding. The case study includes a short description of the problem with a key supporting statement that provides readers with context about the business problem. It is normal in the real world that problems presented are ambiguous, have a small frame of reference, and are based on a number of assumptions.

Short description of the case study

Gamer Moose is a tech start-up that creates a gaming platform for hosting online games. These games can be multiplayer, real-time games hosted on the web and on mobile. It generates money through hosting games provided by different gaming companies and earning a commission on virtual purchases. The application is also available on the App Store and other similar platforms. As it is a start-up, it is competing with similar incumbents, which means deploying incremental features with speed, reliability, and cost-efficiency. In the next few subsections, you will be provided with some assignments that will test your knowledge on the learnings from this book.

How to approach the case study

The presented case study is abstract and it is intended that the readers specifically identify the problem, propose solutions based on assumptions, work out the feasibility of the proposed solution based on assumptions, and finally, make recommendations based on the understanding of the concepts they have learned from this book.

More about tasks

The case study is associated with seven tasks that are presented to readers and expected to test different aspects of continuous delivery in the cloud; these tasks are based on the learnings from various chapters. It is important to note that the authors have provided high-level guidance and possible solutions to get the readers motivated to explore more and come up with a more detailed view of various aspects of problem-solving based on what has been learned from this book.

More about possible solutions

A possible solution only provides the reader with guidance for approaching the problem. It is recommended to explore and apply problem-solving skills along with the knowledge gained from the chapters in this book.

Associated references

The following are references for further reading so readers can gain a more in-depth understanding of the context:

- Game design principles: `https://docs.aws.amazon.com/wellarchitected/latest/games-industry-lens/games-design-principles.htm`
- Game production in the cloud – CI/CD: `https://docs.aws.amazon.com/wellarchitected/latest/games-industry-lens/games-scenario-3.1.html`

Let's move on to the tasks.

Task 1 – business alignment

Your company asks you to develop a strategy for continuous delivery in the cloud. Based on the findings of the strategic section of this book, find out the following:

- Who are the stakeholders of your solution?
- What sources of revenue are there for your delivery strategy and how much revenue is there?
- How can continuous delivery in the cloud support your business?
- How does your business profit from continuous delivery and when do you consider your strategy to be successful?

Guidance toward a possible solution for Task 1

Let's try to address this task with what we learned in *Chapter 3*. To give you an idea of what such a solution could look like, we will go over the planning phase here. This will cover the case study described in the previous section and shed some light on each task.

Business alignment

Gamer Moose is a gaming platform and the majority of its revenue is based on the hosting of the provided games, as well as commissions from in-app purchases. Based on this, Gamer Moose has the following stakeholders:

- **Customers/Gamers**: They expect a stable service, are expecting new features continuously, and don't want to pay too much
- **Vendors**: They want to have a stable platform for their games, expect a fast time-to-market, and don't want to deal with the technical details of the platform
- **App stores**: They expect a secure app for their store

We can expect to become more competitive by providing a good experience, and therefore might be mostly busy with developing the platform. Over time, we expect to get more and more customers, as shown in the following figure:

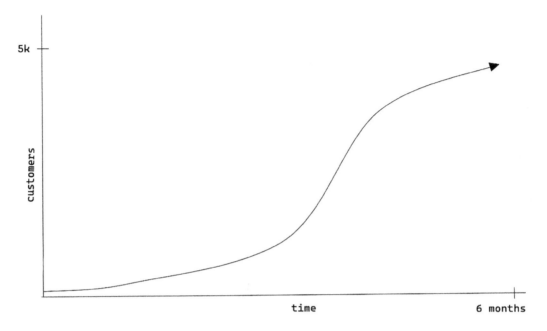

Figure 12.1 – Customers over time

In this case, the use of cloud services can help us in staying cost-effective by scaling the platform based on real-world utilization , but also by keeping the maintenance efforts of our solution low as we might not want to deal with operational topics, at least at the first stage.

Tactical alignment

Before we dig deeper into operational aspects in the next task, it is important to understand the current trends that have disrupted the gaming industry. Essentially, digital platforms have substantially changed the way in which games are created, hosted, and distributed. In this section, we will focus on the tactical aspects of the platform. Let's try to address this using what we learned in *Chapter 4*:

- Shifting left cost, risk, and security management

- Implementation of SRE in the cloud to increase resiliency

- Evolving observability of the overall platform to get a better understanding of potential bottlenecks and gain insights proactively

Task 2 – operational alignment

Your company has limited funding; you are asked to develop a plan for cost-efficient and reliable operations. Based on what you've learned from this book about the operational aspect, find answers to the following:

- How would you implement CD in the cloud to support the business objective?
- Which technology components should be considered to drive efficiency?
- How would you define a high-level organizational structure to drive the operational strategy?

Guidance toward a possible solution for Task 2

Let's try to address this task with the knowledge gained from *Chapters 2* and *4*. To give you an idea of what such a solution could look like, we will go over the planning phase here, which includes the following:

- Choosing the right cloud environment, model, and tool
- Continuous delivery practice and its evolution
- Key practices to adopt and an organizational structure that supports their adoption

As discussed in the *Chapters 2* and *4*, some of the key success factors for aligning business and operations with cost-effectiveness and reliability are as follows:

- Forecasting of infrastructure such as storage, compute and platforms, and resource allocation according to the needs of the applications/services is essential. We have discussed this technique along with specific tools available in the cloud. This would help shift the mindset from owning to consuming the right kind of resources in optimal quantities, reduce the cost of operations through on-demand allocation, and provision these infrastructure capabilities. More information can be found in *Chapter 4* about cost optimization, resource pooling, and provisioning best practices.

- Gamer Moose is a gaming platform. Therefore, operations analytics is an important aspect as well. Adjusting the operational resources for peaks, seasonality, demography, etc. is essential to keep up with the reliability and performance of the platform during busy periods without substantially increasing costs.

- Maintaining live operations for gamers and focusing on the user experience by ensuring the addition of site reliability engineers to the team is also critical.

Task 3 – technical architecture and strategy

The preceding findings will give you an overview of what you might want to consider when building the technical architecture and strategy. Therefore, take the findings from the previous task and define the following things:

- How could a typical stakeholder of your solution benefit from it?
- What are you considering as its success factors?
- Which metrics can be defined to measure this and what could the alternatives be when they are not met?
- Do you want to use one or multiple cloud providers?
- What is your target platform?
- Which deployment strategy do you want to follow?
- Which technologies could help you in achieving your goals?
- How many environments/stages do you need and what purpose do they serve?

Guidance toward a possible solution for Task 3

Let's try to address this task using the learnings from *Chapter 5*. As defined earlier, our business benefits from fast release cycles, low maintenance efforts, and on-demand scaling of the platform. Therefore, we can define the following success factors:

- Short time-to-market from code push to production (a maximum of 24 hours)
- Effective resource utilization (between 80 and 90%)
- Reliability of our platform and CD infrastructure (99.9999%)

By having a short time-to-market, we can ensure that we get very fast feedback on our new features and can react if something goes fundamentally wrong. Given that our platform has to be reliable, we should also consider failure handling and redundancy. Furthermore, our company is a start-up and therefore we should take care with resource utilization. With our strategy, this means that we have to take care of the resource utilization of our infrastructure services, as well as that of the platform.

We also learned that reliability is an issue for our application. Therefore, we might be targeting a multi-cloud strategy in the future, but at the first stage, we could stick with a multi-region setup with one cloud provider. We will come back to this in the final task (*Task 7: Finding the tooling and necessary services to start to implement*).

In this task, let's assume that we are dealing with a microservice application consisting of 10 services that are containerized. Therefore, Kubernetes might be a feasible option, and as we expect the number of services to rise, it is also a future-proof one. In the very first version, we will go with the default rolling upgrade method Kubernetes provides for us. This will be the same for stateful workloads as databases.

As we learned about the benefits of GitOps in this book and want reproducible deployments as well as continuous reconciliation, we will take a GitOps approach to deploy our applications.

Last but not least, we decide that the developers can spin up their lab environments; we are using a quality assurance environment to be able to test migration scenarios of stateful workloads and multiple production environments in different regions, with global load balancers.

Therefore, a very high-level workflow could look like the following:

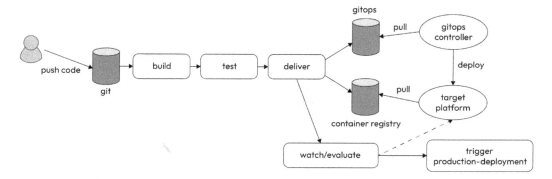

Figure 12.2 – High-level workflow

The preceding figure presents a possible outcome of defining the first workflow from code to production. Keep in mind that the trigger production step, in this case, is only changing a git repository to trigger the deployment on this environment.

Task 4 – getting into the cloud

Now that you have a comprehensive idea of your technical goals, you can take the next step and define how cloud services and technologies could support them. Typical questions to address this are the following:

- What building blocks are there?
- Which service model do you prefer for each building block?
- How much vendor lock-in are you able or willing to accept?
- How many things do you want to build by yourself?

Guidance toward a possible solution for Task 4

Let's try to address this task with what we learned in *Chapter 6*. At this point, we have a kind of vision of what we want to achieve and what it might look like. Everything we created before could work perfectly fine in our own data center, but could incur a large amount of maintenance effort that we as a start-up want to avoid. Given the current architecture, our solution might consist of the following building blocks:

- Source code repository
- Pipeline/workflow tooling
- GitOps controller
- Orchestration platform
- Monitoring

We also decided that we might want to go multi-cloud, which should be taken into consideration when we are selecting the services in *Chapter 11*. At this point, we can define which service models we want to use for each of these building blocks.

For the source code repository as well as the pipeline/workflow tooling and monitoring, we could consider using a SaaS model. Therefore, we won't have to take care of the underlying infrastructure, can utilize runners for the workflows on-demand, and can focus on building our software, whereby such things might already be available as templates on the platform.

To run our workloads, we will utilize a managed Kubernetes environment. Therefore, we don't have to deal with the management of the platform itself and can focus on delivering our applications. At first, our GitOps tooling might run directly on the clusters; in the future, we could consider centralizing them on a separate cluster. As long as it's enough to run containers somewhere, we could also consider using container services from the cloud providers, in which case it is important to know that this could cause some lock-in. Find out what happens if you aren't fast enough, from a technical and business perspective.

Using this approach, we have incurred minimum setup and operations efforts as all of these services are running in a managed way. Nevertheless, we need to have some knowledge about the tools and platforms we are using.

Task 5 – advanced technologies

At this point in time, you have a clear understanding of which services you might need, whether you need to make or buy them, and how cloud providers can support you. In this optional step, you can try to improve your typical deployment strategy and increase your speed by addressing the following:

- What happens if you aren't fast enough, from a technical and business perspective?
- Can feature flagging improve your workflow?

- Do you want to use canary or blue-green deployments and how do you want to deal with them?

Guidance toward a possible solution for Task 5

Let's try to address this task with our learnings from *Chapters 5, 6,* and *7*. Now that we have seen how cloud services can help us, it's time to think about getting faster. As we defined in the beginning and also in our technical strategy, velocity is a crucial factor for the success of our business. For example, a vendor might not be very satisfied if a new game is not published more or less immediately. Furthermore, our users might not perceive our platform as very innovative if nothing changes over the course of several months. To release often, and in a safe way, canary/blue-green deployments and/or feature flags might be the tools of choice.

In our case, feature flags help us to hide new features after they are deployed and gradually enable them for groups of users or individual users. One important thing to remember when using feature flags is the fact that the application code must be instrumented to support this. Another, more infrastructure-centric approach to this is the use of canary or blue-green deployments to ensure that our application is working.

Gamer Moose focuses on finding out whether the services are running in the first step by deploying them in a blue-green way, running some experiments, and shifting the traffic after they find out that the new service is working. In the future, they want to change to canary deployments where they define a group of early adopters, switch them to a new version of the service, and shift them one user group at a time.

Task 6 – security and compliance

Now, you should be at a stage where you know almost every component of your deployment strategy. Find out where there could be security issues and how to mitigate them by addressing the following:

- What potential risks are there?

- What happens if you are not mitigating them?

- What can you do to safeguard your customers?

Guidance toward a possible solution for Task 6

Let's try to address this task with what we learned in *Chapter 8*. Our strategy should contain some security aspects. For our application, we defined reliability as one of the major goals. As we will also deal with billing-related data, confidence might also be an issue and as our app will be deployed on many devices via app stores, integrity is also a major concern.

We've already dealt with the reliability part throughout the previous few sections and confidentiality would probably be a part of application development. Nevertheless, we can ensure the integrity of our software by documenting each step (with reproducible pipelines), signing each artifact, and also validating signatures in each step. As supply-chain security is a large topic on its own, we want to refer to the *Cloud Native Supply Chain Best Practices Whitepaper* (`https://github.com/cncf/tag-security/blob/main/supply-chain-security/supply-chain-security-paper/CNCF_SSCP_v1.pdf`) at this point.

Task 7 – finding the tooling and necessary services to start to implement

Let's try to address this task using our learnings from *Chapters 5, 6, 7*, and *8*. Finally, you can put all these things together, investigate tools and services, and find out which products and services fit your needs. Take into consideration the following questions:

- Are the tools you're using supporting your overall strategy?
- How do the costs of these services/tools relate to the earnings from your application?
- How do these tools play together?

Now, the interesting part of continuous delivery can begin; we are ready to implement our solution. Obviously, we can add additional questions to this catalog that might specifically support our use case. Furthermore, we should take whatever we defined here as law. If we face issues while implementing this, we must rethink our strategy and find out whether there are other ways to implement such things more straightforwardly.

Guidance toward a possible solution for Task 7

Now, we have gotten to the point where we have all of the information needed to make tooling decisions.

For our use case, we should take care of the following criteria:

- How mature is the service?
- Where is it hosted?

Is there a risk of vendor lock-in? One of the decisions we made some time ago was to go multi-cloud, which will influence our tooling decisions. As an example, for container registries and source repositories, we will either select one that is running in multiple clouds on its own, or try to find one that could follow a hub-and-spoke model and push the images to every provider we are using. The same is valid for the monitoring provider. We should ensure that our monitoring is spread across multiple providers and that we will get the relevant information, even if one provider's infrastructure is broken. The CI/workflow tooling is highly dependent on our own preferences. In the case of Gamer Moose, we will take the workflow engine that is attached to the source code management.

Therefore, a selection of services for our strategy could look like the following:

Service	AWS	Google	Azure	Others
SCM (git)	Code Commit	Cloud Source Repositories	Azure Repos	GitHub GitLab
Pipeline	Code Pipeline	Cloud Build	Azure DevOps	TravisCI CircleCI GitHub Actions
Managed Kubernetes	Elastic Kubernetes Service	Google Kubernetes Engine	Azure Kubernetes Service	
GitOps Controllers				ArgoCD Flux

Figure 12.3 – Cloud services

Last but not least, we will use the managed Kubernetes offering that is provided by our cloud service providers. When using multiple providers, we have to take care over their behavior and how we can keep them as similar as possible.

Summary

This chapter serves as guidance for approaching real-world problems. The recommended solution and service options help readers to further enhance their skills. The aim of this chapter is to provide a bird's-eye view approach, not a detailed implementation guide. Finally, it is not possible to provide *off-the-shelf* solutions to such problems; it is suggested to take support from the tools and techniques described in this book and refer to existing projects to enhance understanding.

This book provides a solid foundation for strategizing continuous delivery in the cloud. As the field is growing fast, it requires continuous learning of many aspects of cloud services and the related evolution in the space of continuous delivery. It is recommended that readers continue to look for guidance, best practices, and references as this space matures with time.

Index

www.packtpub.com

Subscribe to our online digital library for full access to over 7,000 books and videos, as well as industry leading tools to help you plan your personal development and advance your career. For more information, please visit our website.

Why subscribe?

- Spend less time learning and more time coding with practical eBooks and Videos from over 4,000 industry professionals

- Improve your learning with Skill Plans built especially for you

- Get a free eBook or video every month

- Fully searchable for easy access to vital information

- Copy and paste, print, and bookmark content

Did you know that Packt offers eBook versions of every book published, with PDF and ePub files available? You can upgrade to the eBook version at www.packtpub.com and as a print book customer, you are entitled to a discount on the eBook copy. Get in touch with us at customercare@packtpub.com for more details.

At www.packtpub.com, you can also read a collection of free technical articles, sign up for a range of free newsletters, and receive exclusive discounts and offers on Packt books and eBooks.

Other Books You May Enjoy

If you enjoyed this book, you may be interested in these other books by Packt:

50 Kubernetes Concepts Every DevOps Engineer Should Know

Michael Levan

ISBN: 9781804611470

- Find out how Kubernetes works on-premises, in the cloud, and in PaaS environments
- Work with networking, cluster management, and application deployment
- Understand why cloud native is crucial for Kubernetes applications
- Deploy apps in different states, including Stateless and Stateful
- Monitor and implement observability in your environment
- Explore the functioning of Kubernetes security at the cluster, user, and application level

Podman for DevOps

Alessandro Arrichiello, Gianni Salinetti

ISBN: 9781803248233

- Understand Podman's daemonless approach as a container engine
- Run, manage, and secure containers with Podman
- Discover the strategies, concepts, and command-line options for using Buildah to build containers from scratch
- Manage OCI images with Skopeo
- Troubleshoot runtime, build, and isolation issues
- Integrate Podman containers with existing networking and system services

Packt is searching for authors like you

If you're interested in becoming an author for Packt, please visit authors.packtpub.com and apply today. We have worked with thousands of developers and tech professionals, just like you, to help them share their insight with the global tech community. You can make a general application, apply for a specific hot topic that we are recruiting an author for, or submit your own idea.

Share your thoughts

Now you've finished *Strategizing Continuous Delivery in the Cloud*, we'd love to hear your thoughts! Scan the QR code below to go straight to the Amazon review page for this book and share your feedback or leave a review on the site that you purchased it from.

https://packt.link/r/1837637539

Your review is important to us and the tech community and will help us make sure we're delivering excellent quality content.

Download a free PDF copy of this book

Thanks for purchasing this book!

Do you like to read on the go but are unable to carry your print books everywhere?

Is your eBook purchase not compatible with the device of your choice?

Don't worry, now with every Packt book you get a DRM-free PDF version of that book at no cost.

Read anywhere, any place, on any device. Search, copy, and paste code from your favorite technical books directly into your application.

The perks don't stop there, you can get exclusive access to discounts, newsletters, and great free content in your inbox daily

Follow these simple steps to get the benefits:

1. Scan the QR code or visit the link below

https://packt.link/free-ebook/9781837637539

2. Submit your proof of purchase
3. That's it! We'll send your free PDF and other benefits to your email directly

www.ingramcontent.com/pod-product-compliance
Lightning Source LLC
Chambersburg PA
CBHW060559060326
40690CB00017B/3759